AF410699

The Numerical Simulation of Fluid Flow

The Numerical Simulation of Fluid Flow

Editors

Robert Castilla
Anton Vernet

MDPI • Basel • Beijing • Wuhan • Barcelona • Belgrade • Manchester • Tokyo • Cluj • Tianjin

Editors
Robert Castilla
Fluid Mechanics
Universitat Politècnica de
Catalunya
Terrassa
Spain

Anton Vernet
Mechanical Engineering
Universitat Rovira i Virgili
Tarragona
Spain

Editorial Office
MDPI
St. Alban-Anlage 66
4052 Basel, Switzerland

This is a reprint of articles from the Special Issue published online in the open access journal *Energies* (ISSN 1996-1073) (available at: www.mdpi.com/journal/energies/special_issues/ Numerical_Simulation_Fluid_Flow).

For citation purposes, cite each article independently as indicated on the article page online and as indicated below:

LastName, A.A.; LastName, B.B.; LastName, C.C. Article Title. *Journal Name* **Year**, *Volume Number*, Page Range.

ISBN 978-3-0365-2931-8 (Hbk)
ISBN 978-3-0365-2930-1 (PDF)

Contents

About the Editors

Robert Castilla

Robert Castilla graduated in Physics from the Universitat Autònoma de Barcelona (Spain) in 1991 and obtained a Ph.D. in Applied Physics from the Universitat Politècnica de Catalunya (Spain) in 2001. He received a position as Associate Professor in the Department of Fluid Mechanics in the Universitat Politècnica de Catalunya, in Terrassa, in 2011. He is also a member of the Steering Committee of CATMech, Center of Mechanical Technologies of Catalonia. His main interest is in Computational Fluid Dynamics with diverse applications: incompressible complex flows (fluid power components as pumps, valves, cylinders, etc.), compressible flows, turbulent flows, aeroacoustics, etc. The present projects are related to supersonic flow in vacuum ejectors and aeroacoustics in cars.

Anton Vernet

Anton Vernet graduated in Chemistry from the Universitat de Barcelona (Spain) in 1991 and obtained a Ph.D. in Chemical Engineering from the Universitat Rovira i Virgili (Spain) in 1997. After one year working as a researcher at the University of Western Ontario (Canada), he obtained a position as associate professor in the area of Fluid Mechanics at the Universitat Rovira i Virgili. He is currently working as a director of the Master in Computational Fluid Mechanics taught jointly between the Universitat Rovira i Virgili and the Universidad Internacional de la Rioja. He is part of the ECoMMFiT research group. His basic lines of research are focused on data analysis, pattern recognition, fuzzy logic and the analysis of images, mainly applied to the study of turbulent flows, heat transfer and biomedical flows. He also works experimentally in the development and application of PIV (Velocimetry by particle imaging) techniques to the study of turbulent flows.

Preface to "The Numerical Simulation of Fluid Flow"

Almost any energy production process involves fluid flow. From the most obvious like air through the wind turbine blades or fuel flow in an internal combustion engine, to a secondary, though still essential, role, like lubrication in a mechanical power transmission hub. In the last decades, the simulation of fluid flows has been so relevant that CFD (Computational Fluid Dynamics) has become a discipline that is included in any textbook of Fluid Mechanics. The growth of computing capacity, summarized in Moore's law, and the development of numerical methodologies provide increasingly efficient and accurate simulations.

This book, which collects the articles published in a Special Issue of *Energies*, aims to be more focused on practical application of available methodologies and models rather than presentation of new numerical methods.

The selected papers for this publication cover a wide variety of topics related to fluid flow numerical simulation. They present problems related to aerodynamics, aeroacoustics and fluid-particle interaction. The ultimate goal has been to show the great diversity of methodologies and procedures that are being used in the numerical solution of flow problems. Some of the papers present innovative methodologies developed for a particular application. Chiu et al. present a solver that allows analyzing the flow-particle interaction. They compare the results obtained with those already published in cases such as Flow Past Circular Cylinders in Tandem or Lid-Driven Semi-Circular Cavity Flow, using them as a benchmark. Gorakifard et al. use the Lattice-Boltzman method to analyze an aeroacoustics problem in wind turbines. They describe the propagation of planar acoustic waves, including the temporal decay of a standing plane wave and the spatial decay of a planar acoustic pulse. The analysis of these specific benchmark problems has yielded qualitative and quantitative data on acoustic dispersion and dissipation, and their deviation from analytical results demonstrates the accuracy of the method. Yuan et al. also address the issue of wind turbines. In this case, they propose a new two-dimensional numerical method, which can facilitate and accelerate the design of the array. Bimbato et al. analyzes a turbulent boundary layer flows past a bluff body and the effect of surface roughness using the Lagrangian vortex method in association with Large-Eddy Simulation (LES). The problem of the effect of roughness is also discussed in the article by de Oliveira al. Xu and Wang use a finite difference (FD) method of monopole source to simulate the response of full-wave acoustic-logging in cave formations. The results of their study show that borehole full-waves have response characteristics to the caves, which are mainly reflected in the energy and the first arrival of waves. Finally, the open-source code OpenFoam is used by Guerrero and Castilla to evaluate the aerodynamic performance of a Formula 1 car under the influence of a turbulent wake.

Robert Castilla, Anton Vernet
Editors

Article

A Fast Two-Dimensional Numerical Method for the Wake Simulation of a Vertical Axis Wind Turbine

Zheng Yuan [1], Jin Jiang [2,*], Jun Zang [3,*], Qihu Sheng [1], Ke Sun [1], Xuewei Zhang [1] and Renwei Ji [1]

[1] College of Shipbuilding Engineering, Harbin Engineering University, Harbin 150001, China; yuanzheng2016@hrbeu.edu.cn (Z.Y.); shengqihu@hrbeu.edu.cn (Q.S.); sunke@hrbeu.edu.cn (K.S.); zhangxuewei@hrbeu.edu.cn (X.Z.); renwei.ji@outlook.com (R.J.)
[2] College of Mechanical and Electrical Engineering, JinLing Institute of Technology, Nanjing 211169, China
[3] Department of Architecture and Civil Engineering, University of Bath, Bath BA2 7AY, UK
* Correspondence: jiangjin@jit.edu.cn (J.J.); j.zang@bath.ac.uk (J.Z.)

Abstract: In the array design of the vertical axis wind turbines (VAWT), the wake effect of the upstream VAWT on the downstream VAWT needs to be considered. In order to simulate the velocity distribution of a VAWT wake rapidly, a new two-dimensional numerical method is proposed, which can make the array design easier and faster. In this new approach, the finite vortex method and vortex particle method are combined to simulate the generation and evolution of the vortex, respectively, the fast multipole method (FMM) is used to accelerate the calculation. Based on a characteristic of the VAWT wake, that is, the velocity distribution can be fitted into a power-law function, a new correction model is introduced to correct the three-dimensional effect of the VAWT wake. Finally, the simulation results can be approximated to the published experimental results in the first-order. As a new numerical method to simulate the complex VAWT wake, this paper proves the feasibility of the method and makes a preliminary validation. This method is not used to simulate the complex three-dimensional turbulent evolution but to simulate the velocity distribution quickly and relatively accurately, which meets the requirement for rapid simulation in the preliminary array design.

Keywords: vertical axis wind turbine (VAWT); two-dimensional wake simulation; finite vortex method; vortex particle method; three-dimensional effect correction model of the wake

Citation: Yuan, Z.; Jiang, J.; Zang, J.; Sheng, Q.; Sun, K.; Zhang, X.; Ji, R. A Fast Two-Dimensional Numerical Method for the Wake Simulation of a Vertical Axis Wind Turbine. *Energies* **2021**, *14*, 49. https://dx.doi.org/10.3390/en14010049

Received: 14 November 2020
Accepted: 20 December 2020
Published: 24 December 2020

Publisher's Note: MDPI stays neutral with regard to jurisdictional claims in published maps and institutional affiliations.

1. Introduction

Since the energy crisis and global warming are getting more serious, the development of renewable energy has received increasing attention. As common renewable energy, wind energy has been well developed due to its wide distribution and nonpollution. A review published by the Intergovernmental Panel on Climate Change (IPCC) shows that the contribution of wind power to the global electricity supply in 2050 will reach 13–14% [1]. There are two common types of wind turbines: horizontal axis wind turbine (HAWT) and vertical axis wind turbine (VAWT). Because of those advantages, such as no need for a yaw control mechanism and low manufacturing costs [2], VAWT has become a research hotspot in recent years. Although the power coefficient of a single VAWT is lower than that of a HAWT, however, some previous research showed that the performance of VAWTs in array configuration would improve significantly. The power coefficient (cp) will increase in a counter-rotating VAWT pairs configuration [3], and the power density of the VAWT farms is potentially more than that of the HAWT farms [4]. To make the VAWT array design more efficient, it is necessary to calculate the velocity field quickly. Therefore, a fast simulation method of the VAWT wake is proposed in this paper so that the wake effect can be measured quickly according to the velocity deficit.

Compared with the HAWT, the study on the wake of a VAWT is still limited. Moreover, in the limited experiments on the wake-field, most of them only focus on the near wake within 3 diameter distances [5–7]. In general, the VAWT array is designed based on the velocity field of the whole wind farm; it is not enough that only near-field data are available, and the complete wake data are necessary. For this reason, there are three requirements that need to be met to be the benchmark test case. These requirements are: First, the velocity profiles from near-field to far-field need to be included in the experimental data (It is better to be as far as 10D downstream). Second, there is no blockage effect; third, the blade is a common type. According to the current research status of the wake experiments, only three experiments [8–10] provide the velocity profiles more than 10D downstream. However, two of them do not meet the other two requirements. In this view, the experiment conducted by WIRE wind tunnel [10] is the only relatively representative experiment so far, which satisfies these three requirements at the same time, and it is chosen as the benchmark test case in this paper. The details of this experiment are clearly reported in Table 2 (Section 3.3).

It should be noted that the scale of the VAWT model in this experiment is not big; the diameter of the rotor is 16.6 cm. However, in order to reduce the blockage effect of the wind tunnel or to obtain relatively farther velocity profiles under the limited measurement distance, many experimental models used in the VAWT wake studies are normally small [11,12]. However, according to the previous studies, the velocity distribution of the wake is quite similar, working on different scale ratios [13,14] and inflow velocities [15,16]. Although the scaling effect of the small model may cause some deviation, in the current, limited research, the benchmark test case selected in this paper is still the most representative. Moreover, only when more experiments of the VAWT wake are published in the future, more validation can be done at that time. However, it wouldn't affect the feasibility of this method in theory and the practicality in engineering.

In this paper, the numerical methods of the VAWT wake simulation are compared and analyzed in Section 2. According to the analysis of each method and the requirement for rapid simulation in array design, the new numerical method is proposed and introduced in detail in Section 3, which includes the finite vortex method, the vortex particle method, the corresponding convergence criteria, and the three-dimensional effect correction of the velocity field. Finally, the numerical results are compared and validated with the published PIV experiment, and the characteristics of the VAWT wake are analyzed in Section 4.

2. Comparison and Analysis of Numerical Methods for Wake Simulation

Because the current wake simulation methods of the vertical axis wind turbine (VAWT) are the same as that of the vertical axis tidal turbine (VATT), therefore, the general numerical methods of the vertical axis turbine (VAT) are analyzed in the following. Apart from the experimental method mentioned above, there are increasing numerical methods to simulate the VAT wake in recent years. Different from the traditional classification of the research methods [17], the simulation process in this paper will be divided into three steps, and each step can be completed by different sub-methods. These three steps and the corresponding sub methods are shown in the Table 1 below:

Table 1. Three steps of the wake simulation and the corresponding sub methods (Note: there is a label in bold before the method, which is the abbreviation of the method and will be used in the following analysis. The number of the label is the step of this process; the letter of the label is the initial of this method).

Main Category	The Description of Corresponding Submethods
The first step: How to calculate the blade load?	**1S**—Streamtube method: The single streamtube model can be used to calculate the performance of the whole rotor, while the multiple streamtube models can be used to calculate the blade load at different azimuthal angles. The influence of upstream on downstream can be considered in the double multiple streamtube model [2]. **1 V**—Vortex Panel method: Based on the Kelvin theorem, the blade forces and vortex shedding at different azimuthal angles can be calculated. According to the types of singularity distribution and the way of the vortex shedding, this method can be further subdivided into many kinds [18–23]. The computational efficiency of the potential theory is high, but the results need to be modified at a high angle of attack [19,24]. **1G**—Grid-based method: There is the body-fitted grid method [25] and the immersed boundary method [26]. Relatively speaking, the computational accuracy is high, but the computational cost is also high. **1 T**—The tabulated data of lift and drag coefficient: It is not necessary to calculate the blade force; only the aerodynamic angle of attack at different azimuthal angles need to be calculated. Then the lift and drag of the blade at the corresponding angle of attack can be obtained by looking up the database table. The lift and drag coefficients were measured at the static state [27] or the dynamic state [24] in wind tunnel experiments. Due to the dependence on the database, it may not be applicable for some new airfoils.
The second step: In which form does the blade force affect the flow field?	**2 F**—The form of force source term: The force coefficient obtained by numerical calculation or database table can be regarded as the force source term, which should be smoothed by the kernel function and calculated by the grid-based method [28]. **2 V**—The form of vorticity source term: The circulation of the bound vortex can be transformed from the force coefficient according to the Kutta–Joukowski law, and the circulation of the shedding vortex can be calculated by the vortex panel method. Then, the circulation will be smoothed into vorticity source terms by the kernel function and calculated by the vortex method [27]. **Note:** This step does not need to be considered in the streamtube method.
The third step: How to simulate the VAT wake?	**3S**—Streamtube method: The streamtube method is based on the momentum theory, and the velocity distribution of the VAT wake can be roughly estimated [2]. **3VE**—Vortex method with empirical viscosity: In order to produce the equivalent effect of viscous dissipation, an empirical decay function is added into the kernel function of the vorticity source term [27]. **3VA**—Vortex method with analytical viscosity: The vorticity source term could be discretized into the vortex particles, and the variation of vorticity caused by the viscous dissipation can be calculated analytically in the vorticity transport equation [29]. **3G**—Grid-based method: The evolution of the wake-field can be simulated by some grid-based methods (such as the finite difference method, the finite volume method [25], the Galerkin finite element method [30] and the lattice Boltzmann method [31]). **Note:** The grid-based method (3G) is in the Eulerian framework, and the vortex method (3VE, 3VA) is a grid-free method in the Lagrangian framework.

The process of the wake simulation can be divided into three steps. In the following examples, the whole simulation process can be represented by the combination of the label mentioned above: Li [27] used the experimental data of static airfoil in the wake research, the evolution of the wake was simulated in the form of the discrete vortex with an empirical decay function (The label of this method is 1T-2V-3VE). The actuator line model of Mendoza [28] used the experimental data modified by the dynamic stall model, and the force coefficient was projected onto the grid to simulate the evolution of the flow field (The label of this method is 1T-2F-3G). There is a hybrid method [32] which was used to calculate the forces on the Euler grid, then the vorticity source term was transformed into the vortex particle in a near-wall transition region, and the far-field was simulated in the Lagrangian framework (The label of this method is 1G-2V-3VA).

Based on the compromise between accuracy and computational efficiency, the vortex panel method was applied to calculate the blade force and vortex shedding because of its affordable computational costs. As a type of the vortex panel method, the finite vortex method [19] is more applicable in a wider range of working conditions due to its improved Kutta condition. However, the simulation of a VAWT wake is a challenging task; the wake evolution of the VAWT is much more complex than that of HAWT, such as the breakup and merging behavior of the vortex swarm. Moreover, there is also a blade-vortex interaction (BVI) problem in the VAWT wake; that is, the blade will encounter the shedding vortex at the downstream half revolution. Nevertheless, compared with the grid-based method, the vortex particle method will be more competent for this job because of its low numerical dissipation and easy-to-parallel. In addition, according to the characteristics of the VAWT wake-field, high vorticity is mainly distributed in the upper and lower rows of the wake instead of the whole wake area. In contrast, the whole domain needs to be calculated in the grid-based method, while only a small region where the vorticity is relatively high needs to be calculated in the vortex particle method. In other words, the calculation domain of the vortex particle method is smaller than that of the grid-based method. Thus, it can be concluded that the evolution of the VAWT wake can be calculated quickly and accurately by the combination of the finite vortex method and the vortex particle method. The corresponding label of this method proposed in this paper is 1 V-2 V-3VA.

To save computing costs, the two-dimensional model was often used in the preliminary array design [33–35]. Even in the atmospheric boundary layer, the wake can also be simplified as the two-dimensional Gaussian wake model [36]. In addition, the aspect ratio (height/diameter) of the VAWT in real engineering practice is usually big, so that more wind energy can be captured by increasing the height without increasing the land area. Therefore, the influence of the three-dimensional effect, which is caused by the blade end, will become smaller in the case of a large aspect ratio. The above reasons make the two-dimensional numerical method more suitable in the VAWT array design.

When the two-dimensional numerical model is used in the preliminary design of the VAWT array, both the blade force and the velocity field should be modified by the three-dimensional effect correction model. There are some three-dimensional effect correction models for the force and performance [37,38], but there are few three-dimensional effect correction models for the VAWT wake. By comparing the three-dimensional and two-dimensional wake simulation [39], it can be found that the length of the three-dimensional VAWT wake is shorter than that of the two-dimensional VAWT wake because velocity deficit will recover in the third dimension. In order to consider the three-dimensional effect in the velocity recovery, a new correction model of the VAWT wake is proposed, which is the projection coefficient from the two-dimensional to three-dimensional along the streamwise direction.

It is true that the turbulent evolution of the VAWT wake is a three-dimensional problem. However, the purpose of this paper is to propose a numerical method that can calculate the velocity distribution quickly and make the VAWT array design more efficient; the complex details of the three-dimensional turbulence evolution can be ignored. Therefore, it is not necessary to pay too much attention to the turbulent problems such as the dynamic stall and the turbulent inflow conditions because these turbulence details are not important for the preliminary array design and will be covered by the three-dimensional effect correction model.

In general, according to the tradeoff between accuracy and efficiency, a two-dimensional numerical method with necessary three-dimensional effect correction is proposed; it can calculate the wake-field quickly and relatively accurately, the first-order approximation of the velocity distribution is sufficient, which meets the requirement in the preliminary array design. More details of this method will be introduced in Section 3.

3. Methodology

The whole process of this new method is like this: First, the finite vortex method is used to calculate the blade load and vortex shedding. Then, the shedding vortex is discretized into the vortex particles to simulate the wake evolution. Finally, it is necessary to introduce a three-dimensional effect correction model into the two-dimensional velocity field so that the modified simulation results can be used in the preliminary array design. Details are given in the following subsections.

3.1. Finite Vortex Method

The forces and circulation of the blade can be calculated accurately by the vortex panel method, which was confirmed in the published papers [18,20]. The finite vortex method is one type of the vortex panel method [19,40]. A schematic of the vertical axis turbine in the reference coordinate is shown in Figure 1. A close-up view of the blade is shown in Figure 2. The motion of the blade can be described by the translational velocity U and angular velocity Ω relative to the local coordinate system o. The inflow velocity is V_A. There are N sources and sinks distributed on the surface of the blade and a linear vortex lineγon the chord line. The schematic diagram is shown in Figure 3.

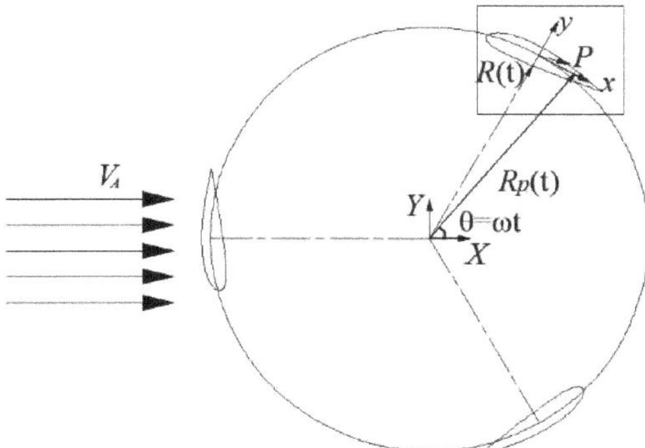

Figure 1. Reference coordinate system.

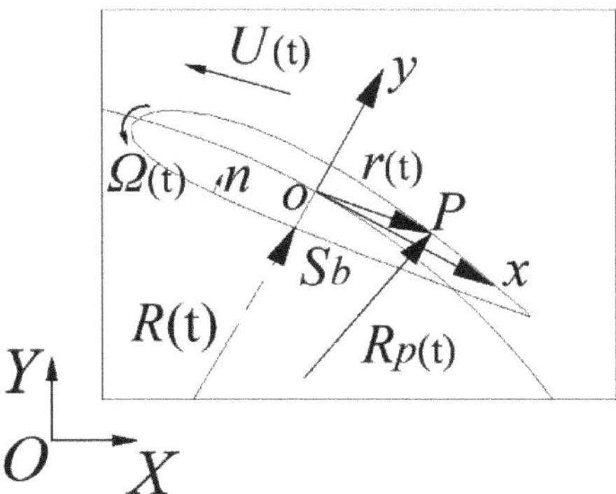

Figure 2. Local coordinate system.

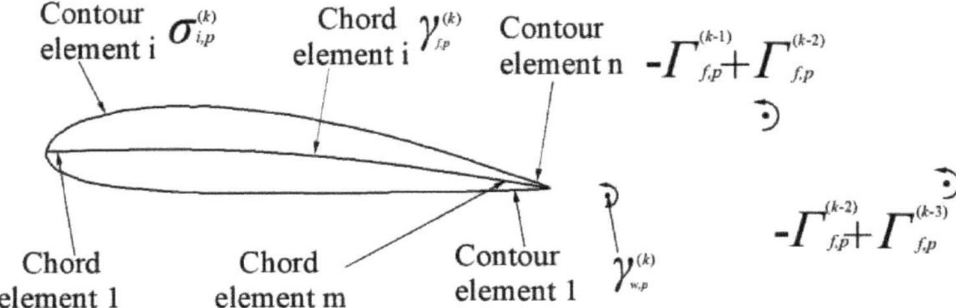

Figure 3. The schematic diagram of the finite vortex method. Subscript i is the serial number of source and sink singularities, p is the number of blades, f represents the attached vortex on the blade, w represents the shedding vortex in the wake, superscript k is the time step.

The total velocity potential ϕ satisfies the Laplace equation in the global coordinate system, as shown in Equation (1) below:

$$\nabla^2 \phi(X, Y, t) = 0 \tag{1}$$

The surface of the blade satisfies no through-flow boundary condition, as shown in Equation (2), where r is the vector from the local coordinate system o to the blade surface, n_b is the unit normal vector that points to the interior of the blade, the symbol S_b denotes the surface of the blade.

$$\frac{\partial \phi}{\partial n}\Big|_{S_b} = \left(\vec{U} - \vec{V_A} + \vec{\Omega} \times \vec{r}\right) \cdot \vec{n_b} \tag{2}$$

The symbol p is the field point; in the far-field, the induced velocity potential φ satisfies $\lim\limits_{p \to \infty} \nabla \varphi = 0$. At the initial time $t = 0$, the induced velocity potential satisfies the relationship shown in Equation (3):

$$\nabla \varphi(p, 0) = 0 \tag{3}$$

Some experimental studies [41,42] on the flow field showed that there is a finite pressure difference at the trailing edge under the unsteady conditions, which makes the streamline roll-up. In order to solve the stall condition at a high angle of attack, the improved Kutta condition set up a finite pressure difference at the trailing edge. Therefore it is called the finite vortex method [19,43]. The subscripts u and d represent the up and downsides of the trailing edge. The improved Kutta condition of the finite vortex model is shown in Equation (4):

$$p_u - p_d = \Delta p \tag{4}$$

In order to determine the value of this pressure difference, set a parameter l to satisfy the relation: $\Delta p = p_u - p_d = -l \rho \gamma_w^{(k)}$, where $\gamma_w^{(k)}$ is the circulation of the vortex shedding at the step k. By adjusting the value of l, the lift coefficient of numerical results would be kept consistent with the lift coefficient obtained by the experiment at different angles of attack [19]. The value of l is close to 0 at the small angle of attack, which indicates that the computational error of conventional Kutta condition (pressure difference at trailing edge is equal to 0) is not large at the small angle of attack. With the increase of angle of attack, the absolute value of l is also increasing, which indicates that the pressure difference of the trailing edge is also increasing gradually. Therefore, due to the existence of the correction factor l and the finite value of pressure difference at the trailing edge, the aerodynamic loads calculated by the finite vortex method can still be consistent with the experimental data at a high angle of attack. It also means that the finite vortex method can be more accurate than the conventional vortex panel method in the case of a high angle of attack.

According to the unsteady Bernoulli equation (as shown in Equation (5)), the pressure in the flow field is:

$$\frac{p - p_\infty}{\rho} = -\frac{\partial \varphi}{\partial t} + \left(\vec{U} - \vec{V_A} + \vec{\Omega} \times \vec{r} \right) \cdot \nabla \varphi - \frac{1}{2}(\nabla \varphi)^2 \tag{5}$$

As the up and downsides of the trailing edge are very close to each other, the vector of the trailing edge can be written as Equation (6):

$$\vec{r_u} \approx \vec{r_d} \approx \frac{1}{2}\left(\vec{r_u} + \vec{r_d} \right) = \vec{r_{TE}} \tag{6}$$

Therefore, after the linearization, the Kutta condition can be rewritten as Equation (7):

$$\begin{aligned}\frac{\partial(\varphi_u - \varphi_d)}{\partial t} = \frac{\partial \Gamma_f}{\partial t} &= \left(\vec{U} - \vec{V_A} + \vec{\Omega} \times \vec{r_{TE}} \right) \cdot (\nabla \varphi_u - \nabla \varphi_d) + \tfrac{1}{2}(\nabla \varphi_d - \nabla \varphi_u) \cdot (\nabla \varphi_d + \nabla \varphi_u) - \frac{\Delta p}{\rho} \\ &= \left(\vec{U} - \vec{V_A} + \vec{\Omega} \times \vec{r_{TE}} - \vec{V_{TE}} \right) \cdot (\nabla \varphi_u - \nabla \varphi_d) - \frac{\Delta p}{\rho}\end{aligned} \tag{7}$$

where mean disturbance velocity vector V_{TE} of the trailing edge is shown in Equation (8):

$$\vec{V_{TE}} = \frac{1}{2}(\nabla \varphi_u + \nabla \varphi_d) \tag{8}$$

According to Kelvin's theorem, the total circulation does not change with time under the potential flow theory; in other words, the sum of circulation of the bound vortex (Γ_f) and free vortex (Γ_w) does not change with time. Hence, there is Equation (9). At different azimuthal angles, the lift and circulation of the airfoil will change. In each time step, the circulation variation of the bound vortex and that of the free vortex are equal in value, but the sign of them are opposite. It can be expressed as Equation (10). Here, the subscript f represents the circulation of the foil, and the subscript w represents the vortex in the wake, the superscript k represents the time step. Combining Equation (10) with Kelvin's theorem (Equation (9)), Equation (11) is obtained. Then, the circulation of the vortex shedding γ can be calculated by inserting Equation (11) into Equation (7) above. Moreover, the γ obtained here is exactly the circulation that will be discretized into the vortex particles. The detailed process is shown in the references [18–20].

$$\frac{d\Gamma_{total}}{dt} = \frac{d\left(\Gamma_f + \Gamma_w \right)}{dt} = 0 \tag{9}$$

$$\Gamma_f^{(k)} - \Gamma_f^{(k-1)} = -\gamma_w^{(k)} \tag{10}$$

$$\frac{d\Gamma_f}{dt} = -\frac{d\Gamma_w}{dt} = \frac{\Gamma_f^{(k)} - \Gamma_f^{(k-1)}}{\Delta t} = -\frac{\gamma_w^{(k)}}{\Delta t} \tag{11}$$

3.2. Vortex Particle Method

The circulation of the shedding vortex is calculated by the finite vortex method in the potential flow. Then, it will be discretized into a vortex particle swarm to calculate the wake evolution in viscous flow. The vortex particle program is divided into six subsections as follows, and the definition of symbols in each equation is only applicable to the relevant subsection.

3.2.1. Discretization of the Circulation

In this paper, the Vatistas model [44] is used as the vortex core model, which is a common smoothing function to remove the singularity, and it is also similar to the Gaussian smoothing function. The Vatistas model also has been applied in the research of a HAWT wake [13]. The regularization function K is shown in Equation (12): (where $\rho = r/r_c$ and r_c is the radius of the vortex core):

$$K(\rho) = \frac{1}{\pi(\rho^4 + 1)^{\frac{3}{2}}} \tag{12}$$

Which satisfies the normalization condition, as shown in Equation (13):

$$2\pi \int K(\rho)\rho d\rho = 1 \tag{13}$$

The function of vorticity distribution is shown in Equation (14):

$$\omega(r) = \frac{\Gamma}{r_c^2}K(\rho) = \frac{\Gamma r_c^4}{\pi(r^4 + r_c^4)^{\frac{3}{2}}} \tag{14}$$

3.2.2. Vorticity Transport Equation

The vortex particle method is a viscous method, the governing equation as follows is the vorticity transport equation, which is obtained by taking the curl of the incompressible NS equation, as shown in Equation (15):

$$\frac{d\omega}{dt} = (\omega\cdot\nabla)V + \nu\nabla^2\omega \tag{15}$$

It is worth mentioning that in the two-dimensional case, there is no stretching term, so the control equation is simplified as Equation (16):

$$\frac{d\omega}{dt} = \nu\nabla^2\omega \tag{16}$$

It can be seen that the governing equation of the vortex particle method in the two-dimensional case is much simpler than the NS equation of the grid-based method.

3.2.3. The Calculation of the Velocity of Vortex Particles

The convection of the vortex particles is calculated by the Biot–Savart Equation. The induced velocity of the Vatistas model in Equation (17):

$$\vec{V} = \frac{\vec{\Gamma}}{2\pi} \times \frac{\vec{r}}{\sqrt{r^4 + r_c^4}} \tag{17}$$

It can be seen that Equation (17) is very close to the induced velocity equation of a two-dimensional singular vortex, as shown in Equation (18):

$$V = \frac{\Gamma}{2\pi r} \tag{18}$$

In this paper, the fast multipole method (FMM) [45] is introduced to accelerate the calculation of the induced velocity of the vortex particles. Here, the induced velocity can be written in the form of induced velocity potential φ, z_0 is the target position, z is the position of the vortex particle, and z_c is the centroid position of the block. The derivation of the Green function is shown in Equation (19):

$$G(z_0, z) = -\frac{1}{2\pi}\log(z_0 - z) = -\frac{1}{2\pi}\left[\log(z_0 - z_c) + \log\left(1 - \frac{z - z_c}{z_0 - z_c}\right)\right] \tag{19}$$

Applying the Taylor series expansion $\log(1 - \xi) = -\sum_{k=1}^{\infty} \frac{\xi^k}{k}$ to the second logarithmic term on the right side of Equation (19), Equation (19) can be transformed into the following form (Equation (20)):

$$G(z_0, z) = \frac{1}{2\pi} \sum_{k=0}^{\infty} O_k(z_0 - z_c) I_k(z - z_c) \tag{20}$$

There are two auxiliary functions ($I_k(z - z_c)$ and $O_k(z_0 - z_c)$) defined here:

$$I_k(z - z_c) = \frac{(z - z_c)^k}{k!}, \text{ for } k \geq 0 \tag{21}$$

$$O_k(z_0 - z_c) = \frac{(k-1)!}{(z_0 - z_c)^k}, \text{ for } k \geq 1; \text{ and } O_0(z_0 - z_c) = -\log(z_0 - z_c) \tag{22}$$

The final expression of velocity potential φ (Equation (23)) and accuracy criterion of order p (Equation (24)) can be written as follows: (Γ is the circulation of each vortex particle, p is the order of Taylor expansion, N is the number of the vortex particles, the error ε is set to 10^{-5} in this paper, R is the diagonal length of the block. Moreover, the black dot here represents the scalar multiplication.)

$$\phi = \frac{1}{2\pi} \sum_{k=1}^{p} \left\{ \frac{(k-1)!}{(z_0 - z_c)^k} \cdot \left[\sum_{0}^{N} \frac{(z - z_c)^k \cdot \Gamma}{k!} \right] \right\} - \frac{1}{2\pi} \ln(z_0 - z_c) \cdot \left[\sum_{0}^{N} \Gamma \right] \tag{23}$$

$$p = \ln \left[\varepsilon \cdot (2\pi) \cdot \left(1 - \frac{R}{|z_0 - z_c|}\right) \right] / \ln\left(\frac{R}{|z_0 - z_c|}\right) - 1 \tag{24}$$

There is an error criterion to judge whether the FMM can replace the direct calculation. The order of the FMM program is *NlogN*; by contrast, the order of the direct calculation method is N^2 [46,47]. In the FMM program, it is necessary to allocate the whole vortex particles into a tree structure and store them in the leaf nodes of the tree structure. The schematic diagram of the fast multipole method is shown in Figure 4.

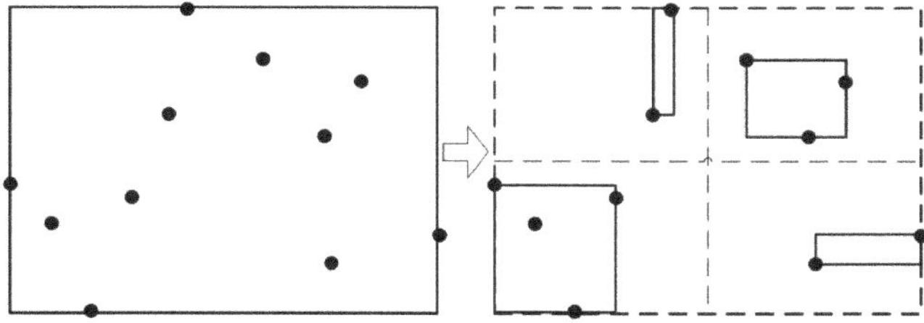

Figure 4. The schematic diagram of how to allocate the vortex particles into the tree structure.

3.2.4. The Calculation of the Vorticity of Vortex Particles

When the position of the vortex particles is updated, it is necessary to update the vorticity of the vortex particles. There are two common methods to simulate the viscous dissipation. One is called the random walk algorithm; the other is the particle strength exchange method (PSE). In this paper, the latter method (PSE) is used. In order to calculate the vorticity dissipation term, a transformation Equation (25) is introduced here to transform the second derivative form of the NS equation into the integral or summation form [48] (The symbol f in the following equation represents the function of vorticity ω).

$$\nabla^2 f(x) = \frac{2}{\sigma^2} \int [f(y) - f(x)] \eta_\sigma(x - y) dy \tag{25}$$

where:

$$\eta_\sigma(r) = \frac{1}{\sigma^2}\eta\left(\frac{r}{\sigma}\right) \text{ and } \eta(\rho) = -\frac{1}{\rho}\frac{dK(\rho)}{d\rho} \tag{26}$$

The governing equation is transformed into the summation form (Equation (27)):

$$\frac{d\omega_P}{dt} = \frac{12v}{\pi}\sum_q \frac{\sigma^4 r_{pq}^2 \left(\Gamma_q - \Gamma_p\right)}{\left(r_{pq}^4 + \sigma^4\right)^{5/2}} \tag{27}$$

Subscript p represents the target point, q is the vortex particles, and the symbol σ is the radius of the vortex particle.

3.2.5. The Calculation of the Turbulent Viscosity Coefficient

It is worth mentioning that the viscosity coefficient v in Equation (27) includes physical viscosity and turbulent viscosity, and the turbulent viscosity is often larger than the physical viscosity. As mentioned in Section 2, in the two-dimensional simulation, no matter what numerical method is used, the three-dimensional effect correction model is needed. However, the turbulence model is still necessary, so here is a brief introduction. The large eddy simulation (LES) in the Lagrangian frame [13] is introduced to calculate the turbulent viscosity in the flow field. The method proposed by Mansfield et al. [49] in Lagrangian large eddy simulation can establish the relationship between filter scale and vortex particle scale. The smoothing function K is related to the particle filter G; the relationship is shown in Equation (28):

$$G_\Delta(r) = K_{\Delta/c}(r) \tag{28}$$

where c is a constant, the filtering scale Δ and the radios of the vortex particle σ satisfy the relationship: $\Delta = c\sigma$. The particle filter and circular filter meet the energy conservation equation in the wavenumber domain. By numerical calculation, the solution $c = \Delta/\sigma = 2.83$ can be obtained. More numerical processes can be found in the work of Liu [13].

When the filtering scale Δ is obtained, the turbulent viscosity can be calculated from the following Equation (29) according to the LES Smagorinsky model [50].

$$\langle v_{sgs} \rangle = (c\Delta)^2 \langle \overline{\omega}\cdot\overline{\omega} \rangle^{1/2} \tag{29}$$

3.2.6. The Redistribution of the Vortex Particles

In order to ensure the convergence of the vortex particle method, a redistribution algorithm is in need [51]. Especially when the distance between particles is too large, the redistribution of particles can ensure a certain overlap rate between particles. There are some interpolation stencils, which distribute the vortex particle on the surrounding nodes with different weights. In this paper, the third-order interpolation function (Equation (30)) is used

$$\Lambda_2(\lambda) = \begin{cases} 1-\lambda^2 & ,0 \le \lambda < 0.5 \\ (\lambda^2 - 3\lambda + 2)/2 & ,0.5 \le \lambda < 1.5 \\ 0 & ,\text{others} \end{cases} \tag{30}$$

$$\Lambda^{2D}\left(\frac{x}{h},\frac{y}{h}\right) = \Lambda\left(\frac{x}{h}\right)\Lambda\left(\frac{y}{h}\right) \tag{31}$$

where $\lambda = x/h$, x is the distance from the center of mesh to the old particles, h is the mesh size. Each old vortex particle will be replaced by nine new particles on the surrounding mesh nodes, as shown in Figure 5.

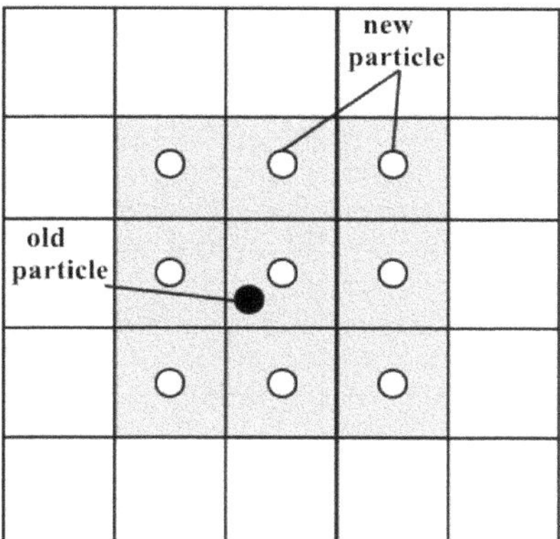

Figure 5. The schematic diagram of the redistribution of one vortex particle.

3.3. Discrete Parameters and Relevant Convergence Criteria

In this paper, the PIV experiment [10] is used as the benchmark test case for validation. The experimental parameters are shown in the Table 2 below.

Table 2. The basic information of the experiment.

Airfoil Section	NACA 0018	Tip Speed Ratio	1.1
Chord length	3 cm	Wind speed at the mid-span	9.43 m/s
Number of blades	3	Chord-based Reynolds number	20,000
Rotor height	15.5 cm	Inflow condition	Turbulent boundary layer
Rotor diameter	16.6 cm	Experimental condition	Wind tunnel, stereo-PIV

The relevant experience of the discrete parameters and the corresponding convergence criteria will be introduced one by one:

The finite vortex method is not a time-consuming algorithm, and in order to ensure convergence accuracy, in this paper, there are 200 discrete elements on the blade surface. The time step of the finite vortex method is the time for $1°$ rotation, and it is a common discrete scale, which is proved by another paper [52]. The numerical parameters in the finite vortex method are conservative, and the accuracy of the program also has been verified by the previously published paper [18].

The diameter of the vortex core in the vortex particle method should be slightly larger than the chord length of VAT. In this case, the diameter of the vortex core is 1.2 times the chord length. This experience is also similar to the selection of the smooth length in the actuator line model [53].

It is worth mentioning that the vortex particle method is a mesh-free method. Therefore, there is no need to verify the independence of mesh size, and it is diffusion free in theory [54]. As long as the number of the vortex particles is enough, the wake evolution could be accurately simulated. The redistribution program in the vortex particle method redistributes each vortex particle to nine new particles through the interpolation function, which satisfies the conservation of the circulation from the zero-order moment to the second-order moment. In this way, the number of the vortex particles will be enough at the place where the vorticity is high, and the vortex particle methods will be automatically adaptive and will concentrate on the regions of physical interest [55]. Since the number of the vortex particles is adjusted adaptively by the redistribution program, the initial number of the vortex particles has little impact on the simulation of the wake-field. Therefore, it is not necessary to verify the independence of the number of vortex particles. In general, the numerical viscosity of the grid-based method will produce a big diffusive error in high Reynolds number. In contrast, the redistribution program does not bring discernible numerical dissipation.

Redistribute the vortex particles in 2D simulations, which shows substantially lower errors than infrequent redistribution [46]. The redistribution program could provide a potential solution for the accurate simulation of the vorticity [56]. It can be seen that the redistribution is of great significance to the convergence of the vortex particle method.

The mesh size of the redistribution program is h = 0.0025 m. In order to ensure that there is a certain overlap rate between adjacent particles, the convergence criterion required that the vortex particle radius should be larger than the mesh size [46]. Hence, the radius of the vortex particle is r_0 = 0.0033 m.

According to the equation $\pi D/c \approx 17.4$, the circumference is 17.4 times of the chord length. Therefore, it is set to discharge the vortex 18 times in one rotation period. That is $360°/18 = 20°$; the time interval of the vortex shedding is the time for $20°$ rotation. The corresponding time step here is 0.0028 s. According to the stability criterion [46], the time step should be less than the reciprocal of the maximum vorticity. In this view, the maximum allowable vorticity under the convergence criterion should be $1/0.0028$ s ≈ 350 s^{-1}. Although the value of vorticity changes with time due to the redistribution, the magnitude of vorticity is within ten in this case, which obviously satisfies the stability criterion.

It can be seen from the benchmark experiment that the blocking effect can be ignored. Therefore, the wake evolution in this paper is simulated under unbounded conditions, and the blocking effect has not been involved. In the vortex particle method, there is no need to set special boundary conditions for unbounded flow. This is a feature different from the grid method.

3.4. The Three-Dimensional Effect Correction Model of Wake

This new correction model is a projection function along the streamwise direction to correct the three-dimensional effect of the velocity field. As we all know, the three-dimensional effect of the wake is that the third dimension will accelerate the recovery of the velocity deficit in the two-dimensional wake. By multiplying the two-dimensional velocity deficit field with a projection function, the influence of the third dimension on the velocity recovery can be equivalent.

The correction process is based on some empirical fitting functions of the VAWT wake. According to the analysis of PIV experiment results [57], the velocity deficit of the wake can be fitted into the power-law function of the downstream position X: $\frac{\Delta U}{U_0} = c_1 \left(\frac{X}{X_t} \right)^{c_2} + c_3$, where X_t is the downstream transition location which is related to the dynamic solidity σ_D, and it can be obtained by this equation: $X_{transition} = D(4.78 - 4.93\sigma_D)$, where D is the diameter of the VAWT. These three coefficients, c_1 c_2 c_3, are the empirical fitting coefficients which are influenced by solidity and tip speed ratio. Moreover, this form of the power-law function is also applicable to the HAWT wake and the two-dimensional or three-dimensional bluff body wake [57].

Therefore, the process of the three-dimensional effect correction proposed in this paper is based on the three-dimensional velocity distribution functions summarized by the previous study [57]. Then, we divided these three-dimensional velocity distribution functions by a two-dimensional velocity distribution function, and the proportion obtained by the division is the projection coefficient from the two-dimensional to the three-dimensional, which is exactly what we want to correct the three-dimensional effect of the wake-field. Finally, the two-dimensional velocity field is multiplied by the projection coefficient along the streamwise direction, and the two-dimensional simulation results considering the three-dimensional effect correction can be used in the preliminary array design.

In order to get the appropriate three-dimensional velocity distribution function which is close to the benchmark case in this paper, two models are selected from the previous study [57] because the parameters of these two models are relatively close to that of the benchmark case, and they were marked as VAWT_1 and VAWT_2, respectively. The velocity distribution of the three-dimensional wake models (VAWT_1 and VAWT_2) and the general form of the velocity distribution of the two-dimensional wake model are described in Table 3 below.

Table 3. The parameters of these three vertical axis wind turbine (VAWT) wake models.

Model	Number of Blades	Tip Speed Ratio	Solidity	The Downstream Transition Location	The Power Function of Wake Velocity Deficit
Benchmark case (two-dimensional wake model)	3	1.1	0.172	$X_t \approx 4D$	$\frac{\Delta U}{U_0} = \left(\frac{X}{X_t}\right)^{-0.5}$
VAWT_1 [57] (three-dimensional wake model)	3	1.2	0.32	$X_t \approx 1.9D$	$\frac{\Delta U}{U_0} = 1.168\left(\frac{X}{X_t}\right)^{-0.684} - 0.098$
VAWT_2 [57] (three-dimensional wake model)	2	1.22	0.21	$X_t \approx 2.88D$	$\frac{\Delta U}{U_0} = 0.542\left(\frac{X}{X_t}\right)^{-0.622} + 0.033$

It can be seen from these functions that the power-law exponent of the two-dimensional wake is -0.5, which is significantly larger than that of the other two three-dimensional wake models (-0.684 and -0.622). That is to say, due to the three-dimensional effect, the decay rate of the two-dimensional wake model is slower than that of the three-dimensional model.

In addition, it should be noted that the two models (VAWT_1 and VAWT_2) selected here are relatively close to the model in this paper; this can be seen in Figure 25 in the reference [57] that there is little difference in the power function between different VAT conditions. Therefore, there is no need to be too concerned about whether the selection of the experimental model is completely consistent with the validation model used in this paper.

In Figure 6, the curves formed by the square symbol (■) represent the power functions of the three-dimensional wake models (VAWT_1 and VAWT_2), the curve formed by the closed circle symbol (•) represents the general form of the power function of the two-dimensional wake model. The functions of these three curves are described above. Then, these two power functions of the three-dimensional wake models (symbol ■) are divided by the power function of the two-dimensional wake model (symbol •), and the ratio functions obtained (symbol ×) are the correction coefficient along the streamwise, which could be used to correct the velocity deficit from the two-dimensional to the three-dimensional.

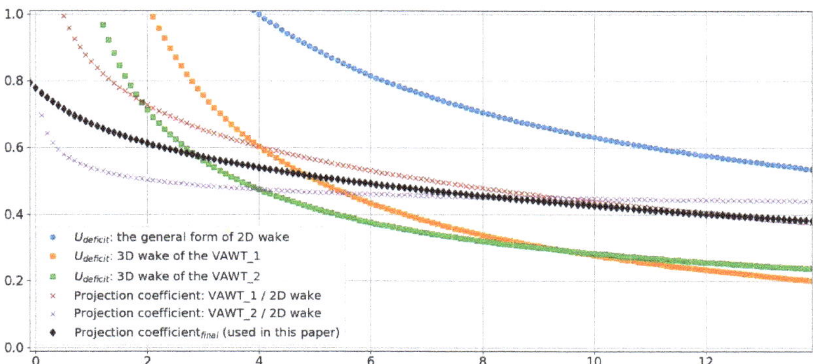

Figure 6. The correction coefficient along the streamwise direction. The X-axis is the downstream position normalized by the rotor diameter (*X/D*). The Y-axis represents two normalized parameters; one is the normalized velocity deficit $\Delta U/U_0$ (symbol • and ■) of the wake model, the other is the correction coefficient (symbol × and ♦) obtained by the division.

There are two other considerations: The first is to eliminate the unreasonable situation that the ratio is much greater than one within one diameter. Second, considering that the three-dimensional effect will make the power coefficient of VAT smaller than that of the two-dimensional model [33], and there is a positive correlation between the power coefficient and the velocity deficit, that is to say, the velocity deficit and power coefficient of the three-dimensional model are smaller than those of the two-dimensional model.

Hence, the correction coefficient should be slightly less than 1 in the near field. Based on these two considerations, combine the two ratio functions (symbol ×), what we get is the empirical coefficient (symbol♦) related to the streamwise position, and the combined function (symbol♦) can be approximately regarded as the final three-dimensional effect projection function which is used in this paper.

It can be found that the curve is a function of the distance along the streamwise direction; by multiplying the two-dimensional velocity field with the projection coefficient (symbol♦) along the streamwise direction, the wake-field corrected by the three-dimensional effect can be obtained. Moreover, this is the whole process of the three-dimensional effect correction for two-dimensional wake simulation in this paper.

In this paper, the power-law function is proposed to correct the three-dimensional effect, which will be a universal idea. Not only the wake of HAWT and VAWT but also the wake of the two-dimensional or three-dimensional bluff body conform to the form of the power-law function [57,58]. Moreover, the two-dimensional Gaussian power function was also used as the basic wake model for array layout design [35,59]. Even if the influence of the boundary layer on the wake is considered, the velocity distribution function could also be fitted into the form of the power-law function [60].

In fact, the correction model is widely used in real engineering practice to simplify the simulation, and it can be equivalent to the effect of some complicated turbulence details, such as dynamic stall and complex turbulent inflow conditions. Especially for the array design, the first-order approximation of the velocity distribution is enough to meet the demand, and the three-dimensional effect correction of the two-dimensional wake simulation is a good choice. The correction model proposed in this paper is derived from the power-law function of velocity distribution, which could make the numerical results in good agreement with the three-dimensional results. The validation of this numerical method is shown in Section 4.

4. Validation and Discussion

In order to verify the reliability of the new numerical method proposed in this paper, a PIV experiment is selected to compare with the simulation results. The parameters of this PIV experiment have been described in detail in Table 2 in Section 3.3.

The comparison results are shown in Figure 7. Relatively speaking, the average velocity profiles calculated by the proposed method match well with the PIV experimental data [10]. At least for the preliminary array design, it is an encouraging result to achieve at least first-order accuracy in the two-dimensional wake simulation. It also proves that the two-dimensional numerical method with a reasonable three-dimensional correction is an acceptable and practicable approach that can make the wake simulation faster and the array design easier. In Figure 8, the rotation direction of the VAWT is counterclockwise. According to the analysis of velocity distribution of a VAWT wake, the peak value of velocity deficit is found not on the centerline. This asymmetry can be seen in the velocity field (Figure 8).

It needs to be clarified here, in the preliminary array design, only the time-averaged velocity field needs to be considered, and the unsteady flow details can be ignored. In addition, since the validation data of the reference is time-averaged, and the benchmark test case did not provide the transient experimental data at different times. In order to keep consistent with the benchmark test case, the time-averaged velocity field was used in this paper.

In the vorticity field (Figure 9), due to the velocity deficit caused by the turbine and the conservation of momentum, the wake vortex will expand in the near field, and the expansion phenomenon of the stream tube in the numerical simulation is similar to the flow field observed by another PIV experiment [5]. It can also be seen that there is a skewness in the distribution of the wake vortex. According to the Magnus effect, a lateral force perpendicular to the inflow direction will be produced on a spinning object. In addition, according to Newton's third law, the forces acting on the turbine and the forces acting on the flow field belong to action and reaction. In this case, the VAWT rotates counterclockwise and is subjected to a downward lateral force. Therefore, the wake is subjected to the upward force and will move upward. The similar phenomena of the wake skewness have also been found in some other VAT wake experiments [8,11]. It can be seen that there is a skewness in the VAT wake caused by the Magnus effect. Moreover, the faster the rotation, the greater the wake skewness. The explanation that wake skewness is related to the Magnus effect can also be found in other studies [5,9,61]. However, from the micro point of view, the generation of the wake skewness is caused by the wind-blade interaction, and it can also be explained from the macro point of view that the wake skewness is caused by the Magnus effect. Both explanations are correct. By comparison, there is usually no skewness in the HAWT wake, this consideration will make the array layout of VAT different from that of HAT. Moreover, more suggestions can be provided for the array arrangement of VATs, such as, the arrangement distance for the downstream VAT should be different considering the rotation direction of the upstream VAT.

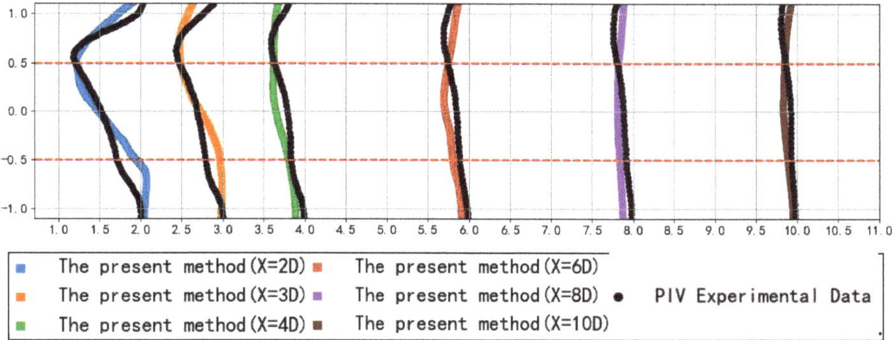

Figure 7. The comparison of average velocity profiles at six different cross-sections. The X-axis represents the sum of the two normalized parameters; one is the streamwise position, which is normalized by the turbine diameter (X/D), the other is the velocity deficit, which is normalized by the inflow velocity ($\Delta U/U_0$). The Y-axis is the lateral position, which is normalized by the turbine diameter (Y/D). The data of black curves are from a published paper [10]; the color curves are calculated by the numerical method proposed in this paper.

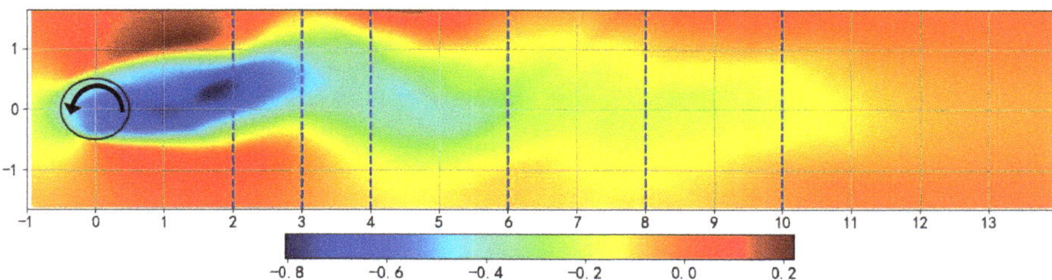

Figure 8. The time-average velocity field of the VAWT wake. The X-axis is the streamwise position, which is normalized by the turbine diameter (X/D). The Y-axis is the lateral position, which is normalized by the turbine diameter (Y/D). The color bar is the velocity deficit, which is normalized by the inflow velocity ($\Delta U/U_0$). The blue dashed lines indicate that the velocity profiles at that location are used for comparison with the PIV experimental results.

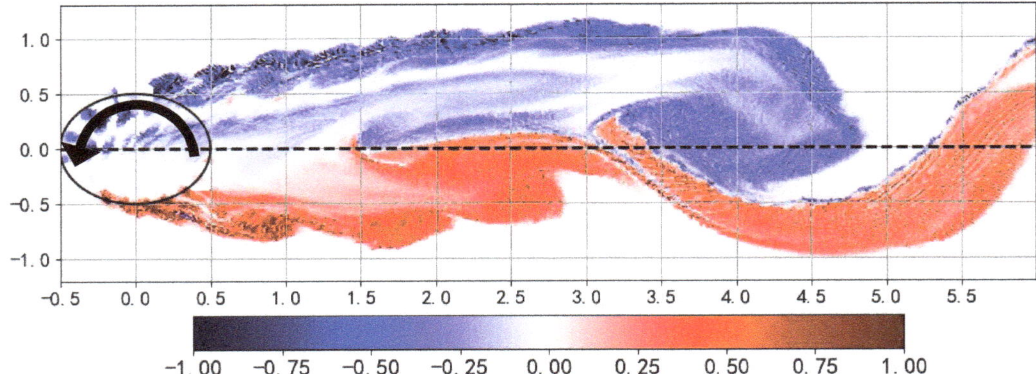

Figure 9. The vorticity field at the near wake. The X-axis is the streamwise position which is normalized by the turbine diameter (*X/D*). The Y-axis is the lateral position which is normalized by the turbine diameter (*Y/D*). The color bar is the nondimensional vorticity which is normalized by the maximum vorticity in the field (ω/ω_{max}).

There is another interesting phenomenon in the far wake in Figure 10. It can be seen from the following eight moments in one period (the period here is not the rotation period, but the period of the vortex street oscillation), due to the complex evolution, the vortex pairs in the far-field are similar to the Karman vortex street in the flow around a cylinder. This characteristic can also be found in another reference [62]. The similar wake evolution will occur in the far-field of a HAT wake as well [39]. It is still unknown whether this complex evolution is similar to the meandering of the HAWT wake. However, at least, this interesting phenomenon in these eight images shown here can lead to a new research direction in the future.

It is also worth mentioning that the vorticity of the vortex particle swarm is different at each time step, and the redistribution program makes the number and vorticity of the vortex particles change all the time. Therefore, the peak value of the vorticity field (ω_{max}) and the range of the color bar (ω) also change at each time step. In order to unify the color bar for observation, the vorticity field (Figures 9 and 10) is normalized by the maximum vorticity at that moment. The color bars of the wake-field are represented by the dimensionless vorticity (ω/ω_{max}). In general, although the redistribution program makes the range of vorticity field change all the time, for the vortex particle method, the whole field still satisfies the conservation condition, so it may not be necessary to pay too much attention to the absolute value of the vorticity at each moment.

However, the purpose of this paper is to propose a fast simulation method for the VAWT wake, so the complex wake evolution phenomenon will not be discussed in depth. The simulation results are calculated by the code written by the authors, and there is no commercial software to use this numerical method so far. In terms of the calculation efficiency of the algorithm, the vortex particle method takes less than 2 h on 4-core CPUs, and the number of the vortex particles is about a hundred thousand, when the focus is only on the flow field within 10D, as we all know, the calculation cost here is much less than that of the grid-based method.

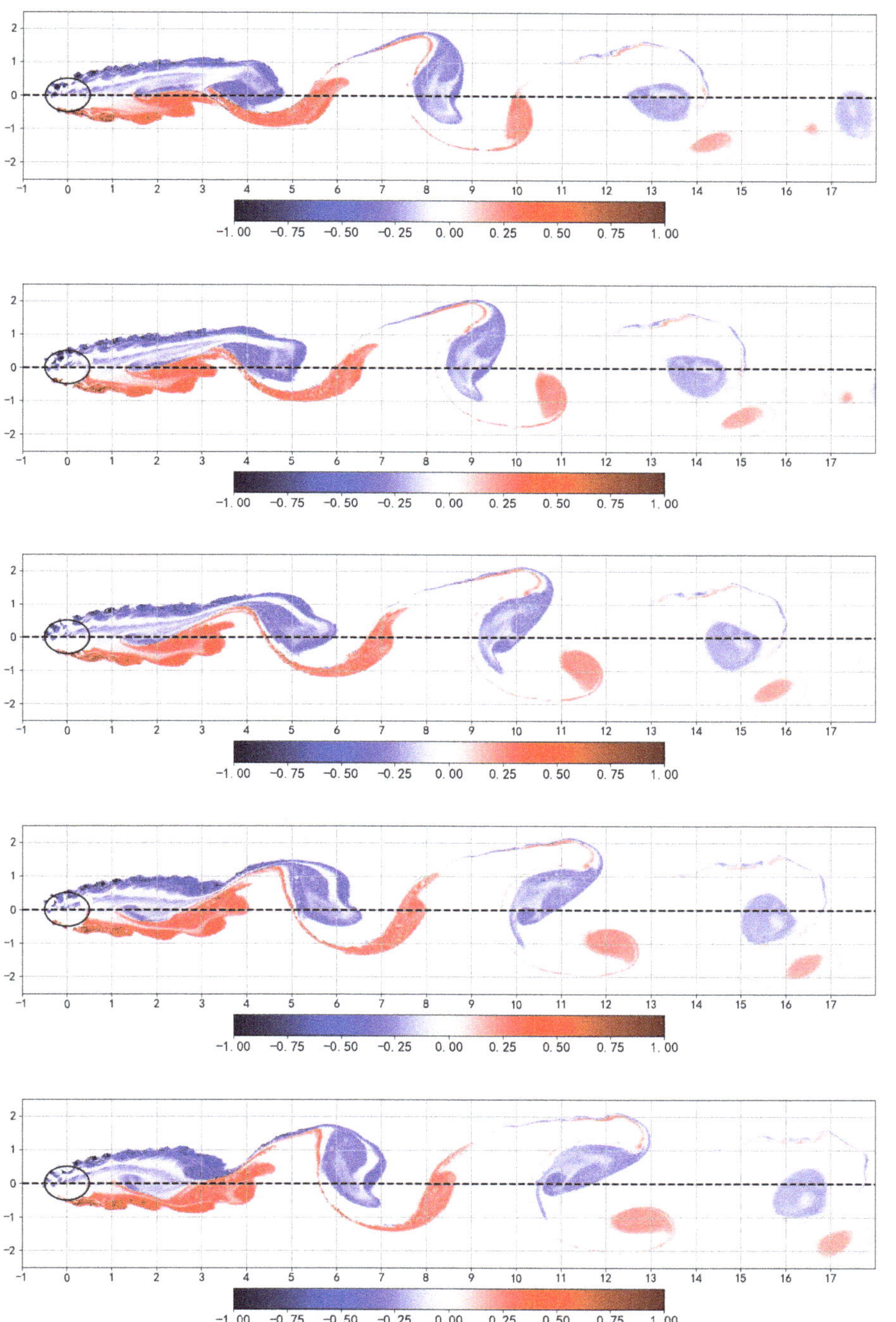

Figure 10. *Cont.*

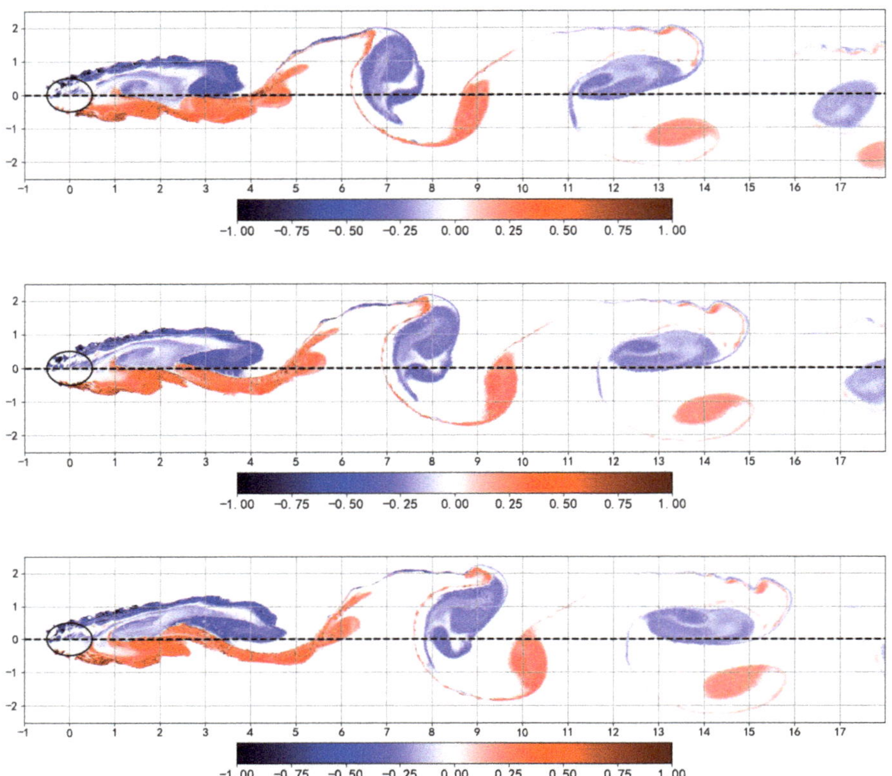

Figure 10. The eight moments of vorticity field in one period at a wide view. The interval of each moment is approximately equal to one rotation period. The X-axis is the streamwise position, which is normalized by the turbine diameter (X/D). The Y-axis is the lateral position, which is normalized by the turbine diameter (Y/D). The color bar is the nondimensional vorticity, which is normalized by the maximum vorticity in the field (ω/ω_{max}).

5. Conclusions

In this paper, by combining the finite vortex method and the vortex particle method, a fast and relatively accurate two-dimensional numerical method is proposed to simulate the wake-field of the vertical axis wind turbine. Based on the power-law function of velocity distribution, a new correction model is proposed to correct the three-dimensional effect of the VAWT wake. Then, the feasibility of the numerical method is preliminarily verified by the PIV experiment. Finally, according to the flow field analysis, the skewness of the VAWT wake is discussed. In the far-field, the VAWT wake evolves into the Karman vortex street, which is similar to the flow around a cylinder.

Although the published experiments on the VAT wake, which can be used for the verification, are relatively limited, however, the numerical method proposed in this paper is feasible in theory and has been preliminarily verified. It has been proven that this new method can simulate the velocity distribution quickly and relatively accurately. The first-order approximation can be achieved by this two-dimensional numerical method, which meets the requirement for efficiency in the preliminary array design. In the future, with more experimental data published, the numerical method proposed in this paper will be systematically studied and improved. Nevertheless, we believe that it can be a promising approach to make the VAWT array design easier and more reasonable.

Author Contributions: Z.Y.: Investigation, methodology, writing the original draft preparation; J.J.: methodology, supervision; J.Z.: methodology, supervision; Q.S.: methodology; K.S.: methodology, validation; X.Z.: methodology; R.J.: Writing—review and editing. All authors have read and agreed to the published version of the manuscript.

Funding: This work was supported by the National Natural Science Foundation of China (No. 51879064, No. U1706227, No. 51979062).

Conflicts of Interest: The authors declare no conflict of interest.

References

1. Wiser, R.; Jenni, K.; Seel, J.; Baker, E.; Hand, M.; Lantz, E.; Smith, A. Expert elicitation survey on future wind energy costs. *Nat. Energy* **2016**, *1*, 16135. [CrossRef]
2. Paraschivoiu, I. *Wind Turbine Design: With Emphasis on Darrieus Concept*; Presses Inter Polytechnique: Montréal, QC, Canada, 2002.
3. Araya, D.B.; Craig, A.E.; Kinzel, M.; Dabiri, J.O. Low-order modeling of wind farm aerodynamics using leaky Rankine bodies. *J. Renew. Sustain. Energy* **2014**, *6*, 063118. [CrossRef]
4. Dabiri, J.O. Potential order-of-magnitude enhancement of wind farm power density via counter-rotating vertical-axis wind turbine arrays. *J. Renew. Sustain. Energy* **2011**, *3*, 043104. [CrossRef]
5. Tescione, G.; Ragni, D.; He, C.; Ferreira, C.S.; Van Bussel, G. Near wake flow analysis of a vertical axis wind turbine by stereoscopic particle image velocimetry. *Renew. Energy* **2014**, *70*, 47–61. [CrossRef]
6. Somoano, M.; Huera-Huarte, F. Flow dynamics inside the rotor of a three straight bladed cross-flow turbine. *Appl. Ocean Res.* **2017**, *69*, 138–147. [CrossRef]
7. Araya, D.B.; Dabiri, J.O. A comparison of wake measurements in motor-driven and flow-driven turbine experiments. *Exp. Fluids* **2015**, *56*, 150. [CrossRef]
8. Peng, H.; Lam, H.-F.; Lee, C. Investigation into the wake aerodynamics of a five-straight-bladed vertical axis wind turbine by wind tunnel tests. *J. Wind. Eng. Ind. Aerodyn.* **2016**, *155*, 23–35. [CrossRef]
9. Ouro, P.; Runge, S.; Luo, Q.; Stoesser, T. Three-dimensionality of the wake recovery behind a vertical axis turbine. *Renew. Energy* **2019**, *133*, 1066–1077. [CrossRef]
10. Rolin, V.F.-C.; Porté-Agel, F. Experimental investigation of vertical-axis wind-turbine wakes in boundary layer flow. *Renew. Energy* **2018**, *118*, 1–13. [CrossRef]
11. Rolin, V.; Porté-Agel, F. Wind-tunnel study of the wake behind a vertical axis wind turbine in a boundary layer flow using stereoscopic particle image velocimetry. *J. Phys. Conf. Ser.* **2015**, *625*, 012012. [CrossRef]
12. Ryan, K.J.; Coletti, F.; Elkins, C.J.; Dabiri, J.O.; Eaton, J.K. Three-dimensional flow field around and downstream of a subscale model rotating vertical axis wind turbine. *Exp. Fluids* **2016**, *57*, 38. [CrossRef]
13. Liu, W.; Liu, W.; Zhang, L.; Sheng, Q.; Zhou, B. A numerical model for wind turbine wakes based on the vortex filament method. *Energy* **2018**, *157*, 561–570. [CrossRef]
14. Bachant, P.; Wosnik, M. Reynolds Number Dependence of Cross-Flow Turbine Performance and Near-Wake Characteristics. APS, A13-006. 2014. Available online: http://vtechworks.lib.vt.edu/handle/10919/49210 (accessed on 25 October 2020).
15. Chamorro, L.P.; Arndt, R.E.A.; Sotiropoulos, F. Reynolds number dependence of turbulence statistics in the wake of wind turbines. *Wind. Energy* **2012**, *15*, 733–742. [CrossRef]
16. Polagye, B.; Cavagnaro, R.; Niblick, A.; Hall, T.; Thomson, J.; Aliseda, A. Cross-flow turbine performance and wake characterization. In Proceedings of the 1st Marine Energy Technology Symposium (METS13), Washington, DC, USA, 10–11 April 2013.
17. Du, L.; Ingram, G.; Dominy, R.G. A review of H-Darrieus wind turbine aerodynamic research. *Proc. Inst. Mech. Eng. Part C J. Mech. Eng. Sci.* **2019**, *233*, 7590–7616. [CrossRef]
18. Wang, L.; Zhang, L.; Zeng, N. A potential flow 2-D vortex panel model: Applications to vertical axis straight blade tidal turbine. *Energy Convers. Manag.* **2007**, *48*, 454–461. [CrossRef]
19. Jiang, J.; Ju, Q.; Yang, Y. Finite Element Vortex Method for Hydrodynamic Analysis of Vertical Axis Cycloidal Tidal Turbine. *J. Coast. Res.* **2019**, *93*, 988–997. [CrossRef]
20. Li, G.; Chen, Q.; Gu, H. An Unsteady Boundary Element Model for Hydrodynamic Performance of a Multi-Blade Vertical-Axis Tidal Turbine. *Water* **2018**, *10*, 1413. [CrossRef]
21. Zou, F.; Riziotis, V.A.; Voutsinas, S.G.; Wang, J. Analysis of vortex-induced and stall-induced vibrations at standstill conditions using a free wake aerodynamic code. *Wind. Energy* **2014**, *18*, 2145–2169. [CrossRef]
22. Abedi, H.; Davidson, L.; Voutsinas, S. Numerical Studies of the Upstream Flow Field Around a Horizontal Axis Wind Turbine. In Proceedings of the 33rd Wind Energy Symposium, Kissimmee, FL, USA, 5–9 January 2015; p. 0495. [CrossRef]
23. Kamemoto, K.; Zhu, B.; Ojima, A. Attractive features of an advanced vortex method and its subjects as a tool of lagrangian LES. In Proceedings of the 14th Japan Society of CFD Symposium, Japan, 1 September 2000; pp. 1–10. Available online: https://www.researchgate.net/publication/237476008_zuixinwofanozhumusubekitexingtoraguranjuLES_fatoshitenoketi_Attractive_Features_of_an_Advanced_Vortex_Method_and_its_Subjects_as_a_Tool_of_Lagrangian_LES (accessed on 25 October 2020).
24. Dyachuk, E.; Goude, A. Numerical Validation of a Vortex Model against ExperimentalData on a Straight-Bladed Vertical Axis Wind Turbine. *Energies* **2015**, *8*, 11800–11820. [CrossRef]

25. Zuo, W.; Wang, X.; Kang, S. Numerical simulations on the wake effect of H-type vertical axis wind turbines. *Energy* **2016**, *106*, 691–700. [CrossRef]
26. Ouro, P.; Stoesser, T. An immersed boundary-based large-eddy simulation approach to predict the performance of vertical axis tidal turbines. *Comput. Fluids* **2017**, *152*, 74–87. [CrossRef]
27. Li, Y. Development of a Procedure for Power Generated from a Tidal Current Turbine Farm. Ph.D. Thesis, University of British Columbia, Vancouver, BC, Canada, 2008.
28. Mendoza, V. Aerodynamic Studies of Vertical Axis Wind Turbines Using the Actuator Line Model. Ph.D. Thesis, Acta Universitatis Upsaliensis, Uppsala, Sweden, 2018.
29. Hu, H.; Gu, B.; Zhang, H.; Song, X.; Zhao, W. Hybrid Vortex Method for the Aerodynamic Analysis of Wind Turbine. *Int. J. Aerosp. Eng.* **2015**, *2015*, 650868. [CrossRef]
30. Nguyen, D.V.; Jansson, J.; Goude, A.; Hoffman, J. Direct Finite Element Simulation of the turbulent flow past a vertical axis wind turbine. *Renew. Energy* **2019**, *135*, 238–247. [CrossRef]
31. Grondeau, M.; Guillou, S.; Mercier, P.; Poizot, E. Wake of a Ducted Vertical Axis Tidal Turbine in Turbulent Flows, LBM Actuator-Line Approach. *Energies* **2019**, *12*, 4273. [CrossRef]
32. Griffith, D.T.; Barone, M.F.; Paquette, J.; Owens, B.C.; Bull, D.L.; Simao-Ferriera, C.; Goupee, A.; Fowler, M. *Design Studies for Deep-Water Floating Offshore Vertical Axis Wind Turbines*; Sandia National Lab. (SNL-NM): Albuquerque, NM, USA, 2018.
33. Li, Y.; Calisal, S.M. Three-dimensional effects and arm effects on modeling a vertical axis tidal current turbine. *Renew. Energy* **2010**, *35*, 2325–2334. [CrossRef]
34. Whittlesey, R.W.; Liska, S.; Dabiri, J.O. Fish schooling as a basis for vertical axis wind turbine farm design. *Bioinspir. Biomim.* **2010**, *5*, 035005. [CrossRef]
35. Gao, X.; Yang, H.; Lu, L. Optimization of wind turbine layout position in a wind farm using a newly-developed two-dimensional wake model. *Appl. Energy* **2016**, *174*, 192–200. [CrossRef]
36. Abkar, M. Theoretical Modeling of Vertical-Axis Wind Turbine Wakes. *Energies* **2018**, *12*, 10. [CrossRef]
37. Zanforlin, S.; DeLuca, S. Effects of the Reynolds number and the tip losses on the optimal aspect ratio of straight-bladed Vertical Axis Wind Turbines. *Energy* **2018**, *148*, 179–195. [CrossRef]
38. Siddiqui, M.S.; Durrani, N.; Akhtar, I. Quantification of the effects of geometric approximations on the performance of a vertical axis wind turbine. *Renew. Energy* **2015**, *74*, 661–670. [CrossRef]
39. Boudreau, M.; Dumas, G. Comparison of the wake recovery of the axial-flow and cross-flow turbine concepts. *J. Wind. Eng. Ind. Aerodyn.* **2017**, *165*, 137–152. [CrossRef]
40. La Mantia, M.; Dabnichki, P. Unsteady panel method for flapping foil. *Eng. Anal. Bound. Elements* **2009**, *33*, 572–580. [CrossRef]
41. Satyanarayana, B.; Davis, S. Experimental Studies of Unsteady Trailing-Edge Conditions. *AIAA J.* **1978**, *16*, 125–129. [CrossRef]
42. Ho, C.M.; Chen, S.H. Unsteady Kutta condition of a plunging airfoil. In *Unsteady Turbulent Shear Flows*; Springer: Berlin/Heidelberg, Germany, 1981; pp. 197–206.
43. Liebe, R. Unsteady flow mechanisms on airfoils: The extended finite vortex model with applications. *Flow Phenom. Nat.* **2007**, *1*, 283–339.
44. Vatistas, G.H.; Kozel, V.; Mih, W.C. A simpler model for concentrated vortices. *Exp. Fluids* **1991**, *11*, 73–76. [CrossRef]
45. Liu, Y. *Fast Multipole Boundary Element Method: Theory and Applications in Engineering*; Cambridge University Press: Cambridge, UK, 2009.
46. Stock, M.J. Summary of Vortex Methods Literature (a Living Document Rife with Opinion). 2007. Available online: http://citeseerx.ist.psu.edu/viewdoc/summary?doi=10.1.1.133.1239 (accessed on 25 October 2020).
47. Winckelmans, G.S. Vortex Methods. In *Encyclopedia of Computational Mechanics*; Chapter 5; American Cancer Society: Atlanta, GA, USA, 2004; Volume 1. Available online: https://onlinelibrary.wiley.com/doi/book/10.1002/0470091355 (accessed on 25 October 2020).
48. Winckelmans, G.; Leonard, A. Contributions to Vortex Particle Methods for the Computation of Three-Dimensional Incompressible Unsteady Flows. *J. Comput. Phys.* **1993**, *109*, 247–273. [CrossRef]
49. Mansfield, J.R.; Knio, O.M.; Meneveau, C. A Dynamic LES Scheme for the Vorticity Transport Equation: Formulation anda PrioriTests. *J. Comput. Phys.* **1998**, *145*, 693–730. [CrossRef]
50. Sagaut, P. *Large Eddy Simulation for Incompressible Flows: An Introduction*; Springer Science & Business Media: New York, NY, USA, 2006.
51. Koumoutsakos, P.D. Direct Numerical Simulations of Unsteady Separated Flows Using Vortex Methods. California Institute of Technology, 1993.1.1. Available online: https://www.researchgate.net/publication/225304763_Direct_numerical_simulations_of_unsteady_separated_flows_using_vortex_methodsPh_D_Thesis (accessed on 25 October 2020).
52. Marsh, P.; Ranmuthugala, D.; Penesis, I.; Thomas, G. The influence of turbulence model and two and three-dimensional domain selection on the simulated performance characteristics of vertical axis tidal turbines. *Renew. Energy* **2017**, *105*, 106–116. [CrossRef]
53. Bachant, P.; Goude, A.; Wosnik, M. Actuator line modeling of vertical-axis turbines. *arXiv* **2016**, arXiv:1605.01449.
54. Papadakis, G. Development of a Hybrid Compressible Vortex Particle Method and Application to External Problems including Helicopter Flows. Ph.D. Thesis, National Technical University of Athens, Athens, Greece, 2014.
55. Beale, J.; Majda, A. Vortex methods. I. Convergence in three dimensions. *Math. Comput.* **1982**, *39*, 1–27.
56. Wang, Y.; Abdel-Maksoud, M. Application of the remeshed vortex method to the simulation of 2D flow around circular cylinder and foil. *Ship Technol. Res.* **2018**, *65*, 79–86. [CrossRef]

57. Araya, D.B.; Colonius, T.; Dabiri, J.O. Transition to bluff-body dynamics in the wake of vertical-axis wind turbines. *J. Fluid Mech.* **2017**, *813*, 346–381. [CrossRef]
58. Tordella, D.; Scarsoglio, S. The first Rcr as a possible measure of the entrainment length in a 2D steady wake. *Phys. Lett. A* **2009**, *373*, 1159–1164. [CrossRef]
59. Wang, L.; Zhou, Y.; Xu, J. Optimal Irregular Wind Farm Design for Continuous Placement of Wind Turbines with a Two-Dimensional Jensen-Gaussian Wake Model. *Appl. Sci.* **2018**, *8*, 2660. [CrossRef]
60. Sun, H.; Yang, H. Numerical investigation of the average wind speed of a single wind turbine and development of a novel three-dimensional multiple wind turbine wake model. *Renew. Energy* **2020**, *147*, 192–203. [CrossRef]
61. Ouro, P.; Stoesser, T. Wake Generated Downstream of a Vertical Axis Tidal Turbine. In Proceedings of the 12th European Wave and Tidal Energy Conference, Cork, Ireland, 27 August–1 September 2017.
62. Zanforlin, S.; Nishino, T. Fluid dynamic mechanisms of enhanced power generation by closely spaced vertical axis wind turbines. *Renew. Energy* **2016**, *99*, 1213–1226. [CrossRef]

Article

Aerodynamic Study of the Wake Effects on a Formula 1 Car

Alex Guerrero and **Robert Castilla** *

LABSON—Department of Fluid Mechanics, Universitat Politècnica de Catalunya, ES-08222 Terrassa, Catalunya, Spain; alex.guerrero.lorente@estudiantat.upc.edu
* Correspondence: robert.castilla@upc.edu

Received: 7 September 2020; Accepted: 23 September 2020; Published: 5 October 2020

Abstract: The high complexity of current Formula One aerodynamics has raised the question of whether an urgent modification in the existing aerodynamic package is required. The present study is based on the evaluation and quantification of the aerodynamic performance on a 2017 spec. adapted Formula 1 car (the latest major aerodynamic update) by means of Computational Fluid Dynamics (CFD) analysis in order to argue whether the 2022 changes in the regulations are justified in terms of aerodynamic necessities. Both free stream and flow disturbance (wake effects) conditions are evaluated in order to study and quantify the effects that the wake may cause on the latter case. The problem is solved by performing different CFD simulations using the OpenFoam solver. The significance and originality of the research may dictate the guidelines towards an overall improvement of the category and it may set a precedent on how to model racing car aerodynamics. The studied behaviour suggests that modern F1 cars are designed and well optimised to run under free stream flows, but they experience drastic aerodynamic losses (ranging from -23% to 62% in downforce coefficients) when running under wake flows. Although the overall aerodynamic loads are reduced, there is a fuel efficiency improvement as the power that is required to overcome the drag is smaller. The modern performance of Ground Effect by means of vortices management represent a very unique and complex way of modelling modern aerodynamics, but at the same time notably compromises the performance of the cars when an overtaking maneuver is intended.

Keywords: Formula 1; Computational Fluid Dynamics (CFD); external aerodynamics; OpenFoam; snappyHexMesh; incompressible flow; Federation Internationale de l'Automobile (FIA); downforce; drag; vortex; wake

---

1. Introduction

For many years, it has been openly stated that F1 has lost much of its spectacular nature due to the difficulty of the cars in being able to follow each other closely for a long period of time. The sophisticated aerodynamics of these single-seater cars has compromised the chasing of the leading car (mostly due to the turbulent wake generation and clearly disturbed flow [1]). This way, it was found to be convenient to numerically analyse and quantify the actual loss of aerodynamic loads on a F1 car due to the immediate presence of a rival in front of it and then, establish a fair comparison with the situation of a free stream condition.

In recent years, a significant number of investigations has been performed, both conducted by the Federation Internationale de l'Automobile (FIA) or Formula One teams and by other sources of investigation. The works of Ravelli and Savini [2,3] have been taken as a reference, as they are considered to be a feasible approach to such a complex CFD problem as the very same CAD (Computer Aided Design) model is evaluated. The obtained results show interesting vorticity behaviours and the reference data for the comparison of their numerical results under free stream flows are taken into account in this manuscript.

Besides, the works of Newbown, et al. [4] and Perry, et al. [5] have set a positive precedent on how to numerically evaluate a F1 car under wake flows, despite using a quite outdated CAD geometry. The methodology that they applied, consisting of varying the different distances between cars, is taken as a good inspiration on how to operate with the present model. However, it is believed that studying closer distances between the cars under the wake flows could potentially give a more complete definition of the study. That is why the closest distance that is evaluated between the two cars on the present manuscript is around a quarter of a car length, which is significantly smaller that the ones that are reflected on other publications.

Other authors, such as Ogawa et al. [6] or Senger et al. [7], have also contributed to the definition of the whole methodology applied this CFD study. Special attention is also placed on the research carried out by Larsson [8] that was backed by the BMW Sauber F1 team, where the description given of the wing interaction with the tire wake is crucial for the understanding of the on-set flow to the underbody.

The main contribution and goal of this paper is to evaluate, study, and numerically quantify the aerodynamic performance of a 2017 spec. adapted F1 car (see Figure 1) under free stream conditions and wake flows with the purpose of being able to argue whether the 2022 changes in the regulations are somehow justified in terms of aerodynamic necessities. In addition to, an original approach is given when including some energetic calculations, so as to see how the fact of running under wake flows can harm or benefit the overall power requirements and, therefore, affect the fuel efficiency of the cars.

This work can be considered to be contemporary study, as very few sources have extracted meaningful conclusions regarding the latest F1 aerodynamic package. Accordingly, the originality of this work recalls on the judgement that the authors may offer on the present F1 regulations, so as to contribute to a significant upcoming improvement of the category.

Figure 1. Ferrari SF70H, a 2017 spec. car. Extracted from Reddit [9].

The CFD methodology is primarily selected as other resources, such as the use of a wind tunnel or any other experimental solutions, are currently out of reach to deal with such a study (see Newbon et al. works in [10]). The amount of energy resources that a wind tunnel experimental execution would require (without mentioning the environmental impact of its construction) is also a fundamental point to opt for a more sensible approach. Moreover, as the CFD discipline involves a rather strict and accurate process to be able to deal with external aerodynamic problems, the methodology is accepted in order to discern and evaluate the hypothesis established. It has to be commented, though, that CFD models are usually calibrated (see [11,12]), but for the present research, it was not possible. Accordingly, a future investigation regarding the usage of experimental data acquisition would validate this approach.

In the following Section 2, the Materials and Methods used along the research are described. In Section 3, the obtained results are presented with a proper discussion and, finally, Section 4 sums up the most relevant and significant conclusions.

2. Materials and Methods

Computational Fluid Dynamics is the branch of Fluid Mechanics that uses numerical methods and algorithms so as to solve and analyse the behaviour of the fluids. In order to benefit from CFD techniques, it is quite important to know as much as possible regarding the real problem that is intended to be simulated (this are the physical properties of the fluid, the boundary conditions as well as other variables). Moreover, CFD involves the design of a CAD model to be meshed and eventually solve a compendium of mathematical equations on it, as they are the fundamental tool to solve in numerical problems. However, some drawbacks that are inherent to CFD involve the model calibration (to ensure reliable results), complex meshing and usually powerful CPU requirements, among others.

Two different scenarios were initially contemplated within the evaluation of the present work. A free stream approach, where the behaviour of the F1 car is analysed under optimal conditions and a more challenging one, based on a conditioned situation under wake flows. As for the latter, different distances were evaluated in order to appreciate the changes and behaviour of the car in second place in direct comparison with the leading car.

2.1. Geometry

The CAD geometry employed was the PERRIN F1 car [13], as it was a realistic approach to a real F1 car because it was designed under the 2017 FIA regulations. This includes several small realistic details, such as winglets, vanes, vorticity generators and slots, being the smallest elements 1.5 mm thick. The length L of the car, which is later considered for the dimensions of the overall domain is around 5.350 m, while its wheelbase, required for the turbulent length scale measures 3.745 m. The geometry was exported to Parasolid format (.*x_t*.), so as to be processed by the SnappyHexMesh command. Figure 2 displays a generic view of the aforementioned F1 car.

Figure 2. Generic view of the F1 car CAD model used.

2.2. Solver

The simulations were executed using the OpenFoam [14] toolbox that solves the equations for unsteady incompressible flow of mass and momentum presented in Equation (1), (also known as Navier–Stokes equations [15]); as well as the continuity equation shown in (3).

$$\rho\frac{\partial \mathbf{u}}{\partial t} + \rho(\mathbf{u}\cdot\nabla)\mathbf{u} = \rho\mathbf{g} - \nabla p + \mu\triangle\mathbf{u} \tag{1}$$

where

$$\triangle \equiv \nabla^2 = \frac{\partial^2}{\partial x^2} + \frac{\partial^2}{\partial y^2} + \frac{\partial^2}{\partial z^2} \tag{2}$$

$$\nabla\mathbf{u} = 0 \tag{3}$$

The discretisation of the domain was obtained applying the FVM (Finite Volume Method), as this method is locally conservative due to being based on a balance approach [16]. The gradient, divergence, and laplacian terms of the Navier–Stokes equations were discretised by means of the Gaussian schemes: at cell interfaces, the interpolation schemes were linear (upwind).

On the other hand, the simpleFoam algorithm (Semi-Implicit Method for Pressure-Linked Equations) was the one chosen, as it is appropriate for incompressible, steady, turbulent flows. The reason why a steady solver was applied on a flow field with unsteady characteristics resides in the nature of turbulence, which is, by definition, non-stationary. Accordingly, the presented results (wake vortices among others) are averaged on a medium flow, which is what simpleFoam solves. The goal was not to study transient phenomena (which could be the subject for future works), but these averages to evaluate the effect on average stability of the car.

The GAMG (Geometric Algebraic Multi Grid) solver was utilised for the pressure equation, while smoothSolver was the one that was selected for velocity and turbulence variables.

As for the execution, the initial 800 iterations were performed under a first order discretisation, while the remaining ones were performed using second order schemes.

For the turbulence model, the k-ω SST (k-ϵ is used in the outer region of and outside of the boundary layer and k-ω is used in the inner boundary layer [17]) was selected, since it offered a faster convergence and better results as compared to k-ϵ and Spart Allmaras.

2.3. Domain and Mesh

The fluid domain dimensions were inspired by other publications, such as Broniszewski's [18], who puts special attention to the back region of the domain in order to be able to capture the generation of the wake properly. Figure 3 shows the boundaries of the problem.

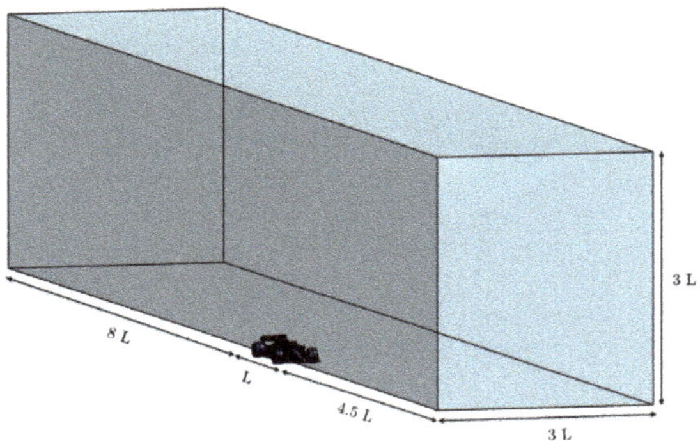

Figure 3. Fluid domain dimensions.

A three dimensional hexahedral mesh by means of the snappyHexMesh and blockMesh commands was generated. The height of the first cell at the solid surfaces is set at 0.01 mm with a layer expansion ratio of 1.3. The resulting average value of y+ is around 40, which involves the use of wall functions as opposed to a near wall treatment, which generally adopts y+ around 1 when solving for low Reynolds (Re) number [19].

Many refinement enclosures were specified along the geometry in order to be able to capture the effects of the wake and and other phenomena. Special attention was placed in the massive wake region as well as on the downforce generation zones, such as the wings and floor. The results of the mesh procedures may be checked in Figures 4–6.

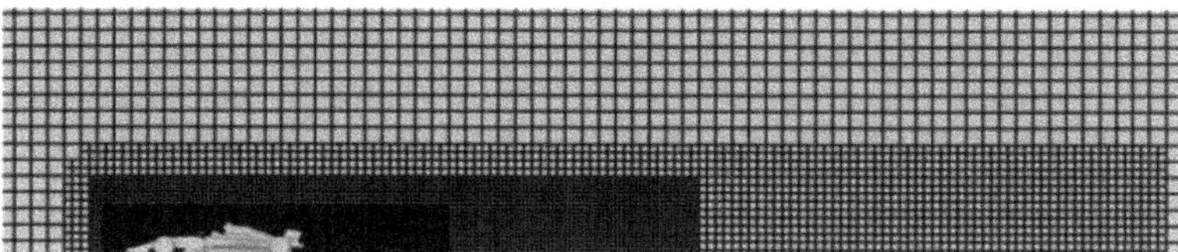

Figure 4. Overall mesh and refinement enclosures of the free stream case.

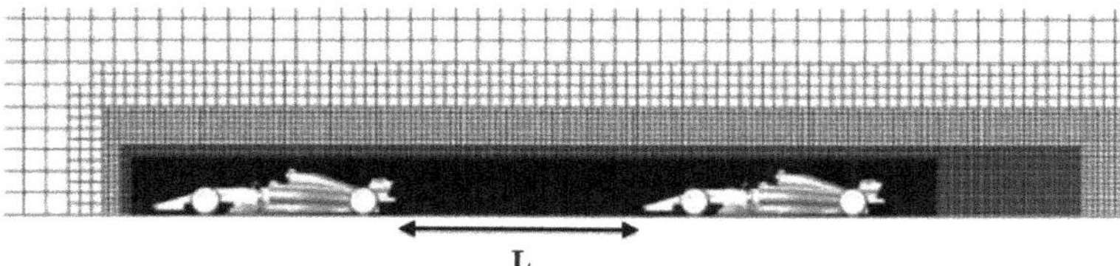

Figure 5. Overall mesh and refinement enclosures under wake effects.

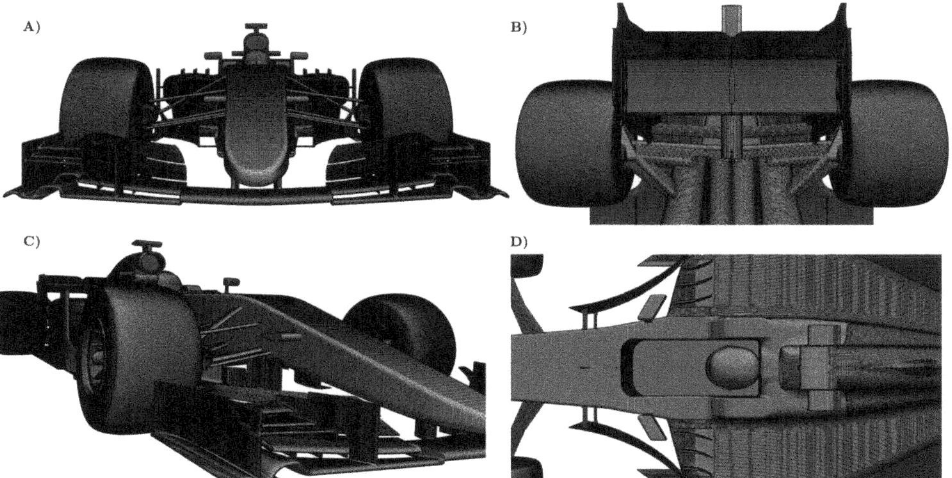

Figure 6. Mesh (**A**) front, (**B**) rear wing, (**C**) trimetric, and (**D**) cockpit.

A Grid Convergence Index (GCI) has been calculated in order to guarantee that the analysis of the results that were obtained was independent of the grid size. The GCI methodology is performed according to Celik et al. [20] and the study is performed by varying the level of definition of the three different meshes studied. The refinement levels used range from 5–6 in the coarsest, 6–7 in the intermediate, and 7–8 in the finest. The ϕ value analysed corresponds to the lift coefficient multiplied by the surface ($S \cdot C_L$). Table 1 presents the results obtained of the GCI.

Table 1. Grid Convergence Index result of the meshes studied.

Mesh Parameters	$\phi = S{\cdot}C_L$
Number of elements N1, N2, N3	18055055, 13793013, 10634551
Refinement ratio r_{21}	1.309
Refinement ratio r_{32}	1.297
Finest result ϕ_1	−3.43
Intermediate result ϕ_2	−3.40
Coarsest result ϕ_3	−3.28
Order of convergence p	4.37
Error GCI_{fine}	1.1%
Error GCI_{coarse}	3.7%

The GCI values are in the asymptotic range of convergence, both GCI_{fine} and GCI_{coarse}, as shown in Table 1. It is as well possible to note that the difference in ϕ values between the finest and intermediate meshes is considerably small thus, obtaining a rather small percentage of error. As for the computational resources, the finest mesh (which is formed by 19.2 million cells in the free stream case and around 40 million under wake flows) took an approximated time of 8 and 14 hours respectively to be generated. The grid size is found to be reasonable as it is almost identical to the one with which the results are later compared to (see Section 3). Hence, as computational resources do not suppose a huge compromise nor restriction, it was preferred to choose the finest mesh for the following study. The sacrifice in computational time and consumption was notably overcome by the superior amount of detail and definition that the finest mesh could deliver.

2.4. Boundary Conditions

The general boundary conditions established are as follows:

- Inlet velocity set at 50 m/s, as it corresponds to a value which can be easily replicated in a wind tunnel (for experimental validation purposes).
- Pressure outlet set at Atmospheric pressure.
- Symmetry plane.
- Ground velocity set at 50 m/s.
- Slip condition on the side wall and the top of the fluid domain.
- Angular velocity and rotational axis of the wheels (MRF).

According to [21], it is possible to assign the car wheelbase, which represents the size of the largest eddy, as turbulent length scale *l*. All of these data are shown in Table 2.

Table 2. Setup of the simulation.

Variable	Value
Free stream velocity U_∞	50 m/s
Fluid density (ρ)	1.225 kg/m^3
Turbulent Intensity (I) [22]	0.15%
Turbulent length scale (l)	3.475 m
Reynolds Number (Re)	12×10^6

2.5. Simulation Performance

The simulations were executed in a cluster that was equipped with 32 cores and 64 Gb of RAM memory and the post-processing operations were carried out by means of an i7 six core laptop with 16 Gb of RAM.

Finally, it was considered to reach convergence when the pressure and velocity residuals were lower than 10^{-3}, just as seen in Figure 7. The run time of the cases was around 10 hours under free stream conditions and 17 h under wake flows.

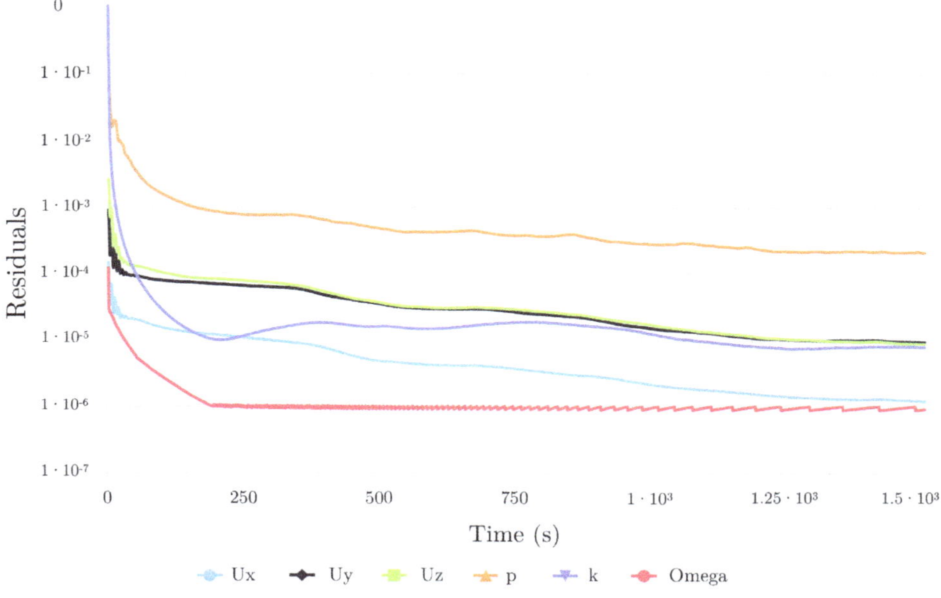

Figure 7. Residuals of the simulation

The obtained results are compared with the reference data proposed by Perrin [23], which are obtained through a CFD simulation using TotalSim, a professional enterprise highly experienced in CFD problems. This is taken as a reference as the boundary conditions and the problem definition coincide with the ones that are proposed in this study, although the solvers are different. Besides, reference data have a public background of more than 120 runs evaluated, so this may be considered to be a reliable source of validation.

The numerical prediction deals with the aerodynamic parameters: Downforce (SC_L), Drag (SC_D), overall aerodynamic efficiency E (C_L/C_D), and Front Balance FB, which is defined as the fraction between the downforce that is generated by the front axle and the total one.

3. Results and Discussion

3.1. Free Stream Condition

Table 3 shows that the results of the performed RANS simulation with k-ωSST model encounter a notable agreement with the reference data. The error found in both downforce and drag coefficient differs by 4.45% and 6.50%, respectively. This indicates that the overall simulation is acceptable, as the difference between the aerodynamic coefficients is found to be small enough to accept them and validate the reference datum.

Table 3. Comparison between the simulation and reference data.

	SC_L (m^2)	SC_D (m^2)	$\lvert C_L/C_D \rvert$	FB (%)
Reference Data	−3.59	1.23	2.92	44.80
Simulation Results	−3.43	1.15	2.98	41.23
Error (%)	4.45	6.50	2.15	7.97

On the other hand, it is found appropriate to indicate the relative contribution in terms of downforce and drag of the main components of the car. This way, it is possible to see the influence of each element and appreciate its aerodynamic efficiency, which is very useful in terms of redesign purposes and issues detection. The different components evaluated can be checked in Figure 8, while Table 4 shows the distribution of these aerodynamic forces.

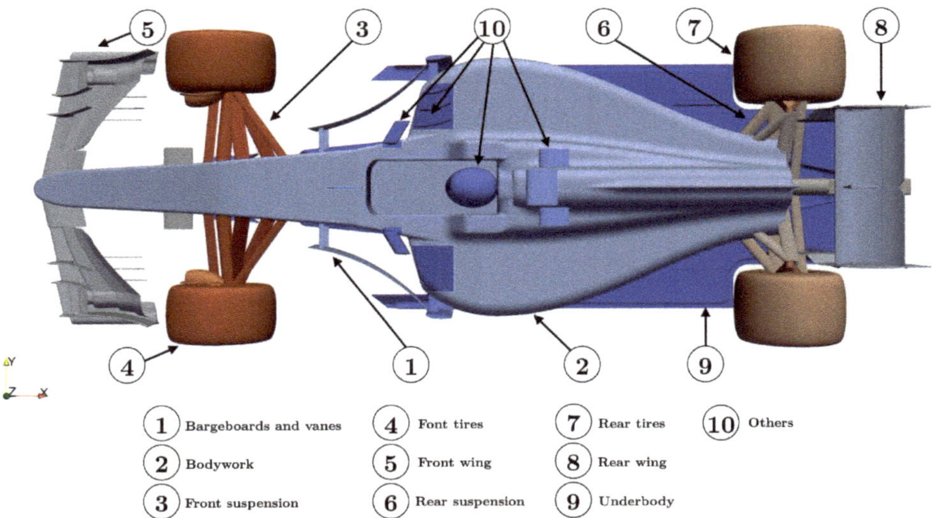

Figure 8. Evaluated components of a F1.

Table 4. Relative downforce and drag contributions of the different elements.

Elements	Downforce Contribution (%)	Drag Contribution (%)
Bargeboards and vanes	−1.71	+2.97
Bodywork	+21.10	+10.7
Front suspension	+2.36	+2.88
Front Tires	+2.94	+10.30
Front Wing	−22.84	+12.69
Rear suspension	−0.40	+3.93
Rear tires	+1.42	+20.06
Rear wing	−35.46	+20.38
Underbody	−61.09	+15.05
Others	−6.65	+1

The underbody, which is composed by the flat floor, the plank, and the diffuser, is responsible for the 60% of the total downforce generation. Following this trend, the rear and the front wing represent, respectively, around 35% and 23% of the overall downforce of the car. The bodywork, shaped as a wing profile, counterbalances these gains by producing lift as well as other elements, such as the front suspension, and both rear and front tires.

On the other hand, Table 4 shows that tires represent around 30% of the total drag, specially on rear's, as the wheels are not covered. This area is well influenced and dominated by big pressure losses in the wake, which leads to these undesired highly turbulent zones of the total drag of the car. Additionally, the front and the rear wing are also present, by being responsible for 13% and 20% of the total drag, respectively. Other elements, such as the underbody (15% of the drag) and the bodywork (10.70%), take an important contribution regarding the aerodynamic resistance.

In terms of aerodynamic efficiency, it is important to note that the underbody is, with no hesitation, the most efficient part of the car. This can be explained due to the use of Ground effect (despite being limited by the flat floor and the size of the diffuser). As opposed to that, the rear wing is known for its low aspect ratio, which helps to generate big downforce quantities, but it suffers from the production of induced drag. The efficiency of the front wing is somewhere between the underbody and the rear wing: the beneficial points of the rake and ground effect that are commented in [24] are counterbalanced by the high angle of attack of the flaps and the vortices generated at the tips.

Figure 9 shows the pressure distribution by means of the dimensionless coefficient (C_P) around the car, both in terms of a plot and in a three dimensional visualisation. The upper view reflects high-pressure zones that are located in the nose (specially in the front wing) as well the rear wing due to its high curvature. Some stagnation areas are also found around the cockpit, where the pressure distribution is low and smooth. These results are in line with what it was previously commented, with the wings being one of the main generators of downforce due to its high angle of attack and curvature.

On the other hand, the underbody of the car shows low-pressure zones under the wings (as it was clearly expected by its nature and shape). Besides, the low-pressure zones along the floor and diffuser suggest that the car is working properly under the Ground effect. It is possible to see a smooth transition from a medium pressure zone to a low pressure region (meaning that the airflow is being accelerated), and finally an increase of pressure that returns the airflow in a lower velocity to the wake. However, the region in close proximity to the tires is affected by an increase of pressure, which implies the loss of the Ground effect benefits.

In general, the pressure distribution is rather smooth around the car, with no abrupt pressure gradients or unexpected transitions, just like that observed in [4].

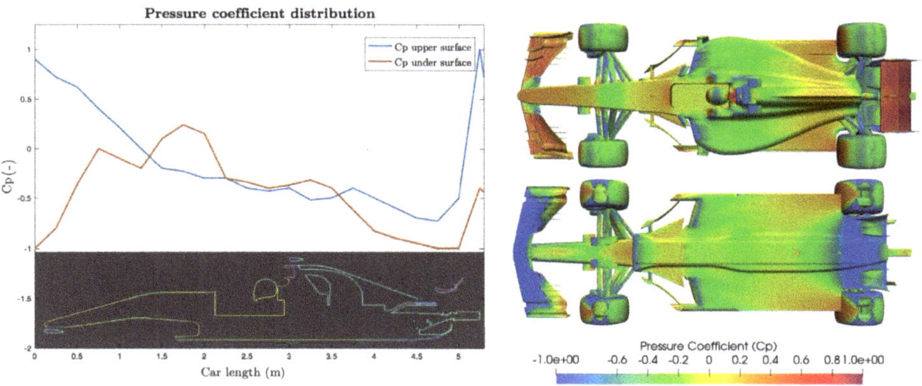

Figure 9. Pressure coefficient plot and representation on the upper and under surfaces of the car.

Figure 10 pretends to illustrate to the reader the shape, position and prolongation of some three-dimensional vortical structures, as well as provide several details regarding the strength and rotational axis of such vortices.

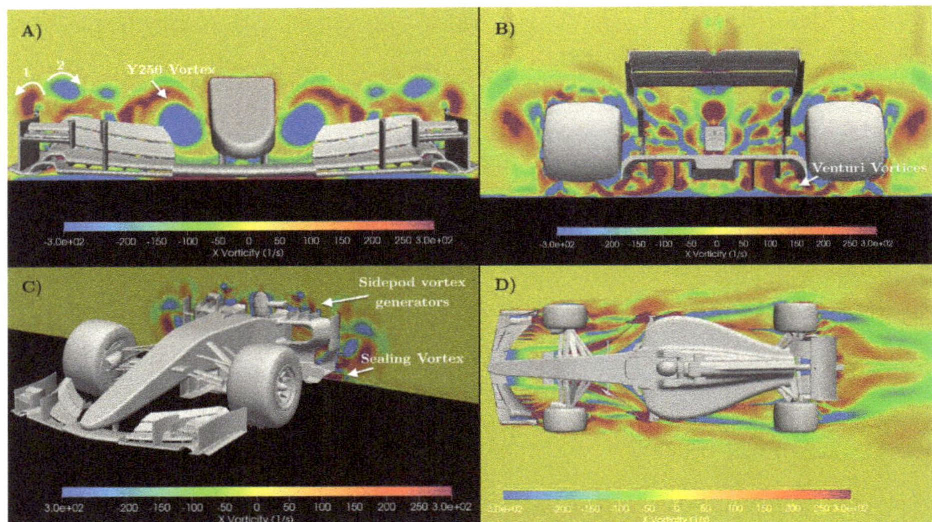

Figure 10. Several axial vorticity contours: (**A**) YZ plane at 0.65 m in the X direction, (**B**) YZ plane at 3.65 m in the X direction, (**C**) YZ plane at 2.60 m in the X direction, and (**D**) XY plane at 0 m in the Z direction.

Figure 10A shows the multifaceted front wing: not only generating downforce, but also axial vorticity that tends to avoid the front tires and energises the flow downstream. Several vortical structures can be seen on the tip of the endplate (A1) or the winglet endplate (A2), among others. The rotation of the vortices—clockwise or counter clockwise—depends on the pressure field around them [25]. A quite interesting phenomena, as it is clearly the biggest vortex generated in the front wing, is the so-called Y250 vortex. This vortex is developed between the middle section of the wing and the multi flap surface, and it is aimed at recirculating the flow towards the underbody of the car (*inwash*).

Figure 10B, shows the rear part of the car, where several vortices are originated as a result of the wake of the spinning wheels and other devices. Special attention is placed in the Venturi vortices that are generated on the side of the diffuser due to the pressure gradient between the underbody and the outside. Additionally, the strakes of the diffuser generate small vortices that are coupled with an opposite rotating vortex due to the interaction with the ground boundary layer, similar to the ones observed in [26].

Figure 10C reflects the generation of vortices in the upper middle region of the car. The bargeboards and vanes play a special and important role here. The goal of the latter is to seal and canalise the flow over the bodywork, making sure that the flow keeps attached along the car (Coanda effect). Besides, the sealing vortices that are generated by the bargeboards tend to act as skirts, therefore preventing the underbody airflow to escape and maximise the Ground effect [27] and the diffuser efficiency.

Figure 10D displays a top view of the whole car to understand the behaviour of the overall vorticity.

Finally, Figure 11 shows a three dimensional representation of the streamlines of the vorticity. It is possible to see, in general traits, how the flow behaves around different areas of the car (Y250 vortex, tip of the wings, middle section) and how chaotic and turbulent the resulting wake looks like.

Figure 11. Streamlines of the axial vorticity.

3.2. Under Wake flows

Similarly, the evaluated results deal with the aerodynamic parameters: downforce (SC_L), drag (SC_D), overall aerodynamic efficiency E (C_L/C_D), and front balance FB.

However, four different situations are studied, which differ in the distance established between the 2 cars: 0.25 L, 0.5 L, 1 L, and 2 L, where L represents the length of one car (approximately 5.3 m). Figures 12–15 display the evaluated parameters of the second car (follower car) and Table 5 shows the percentage of change of those with respect to the first car (leading car).

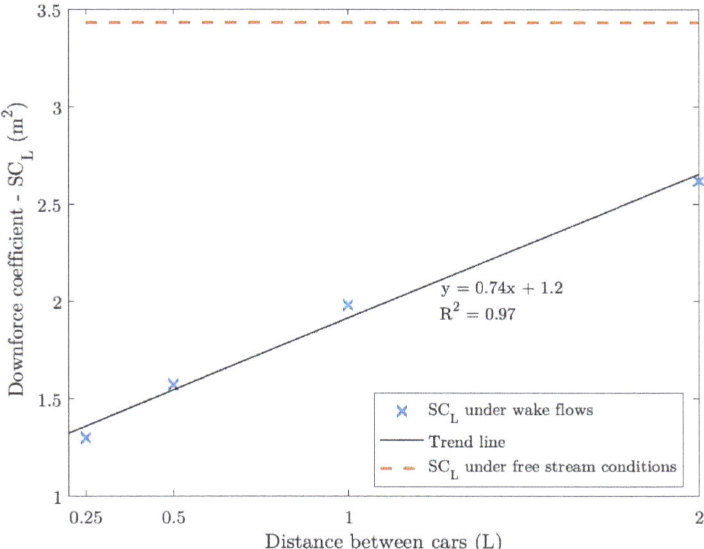

Figure 12. Comparison between values of SC_L under wake flows and under free stream conditions.

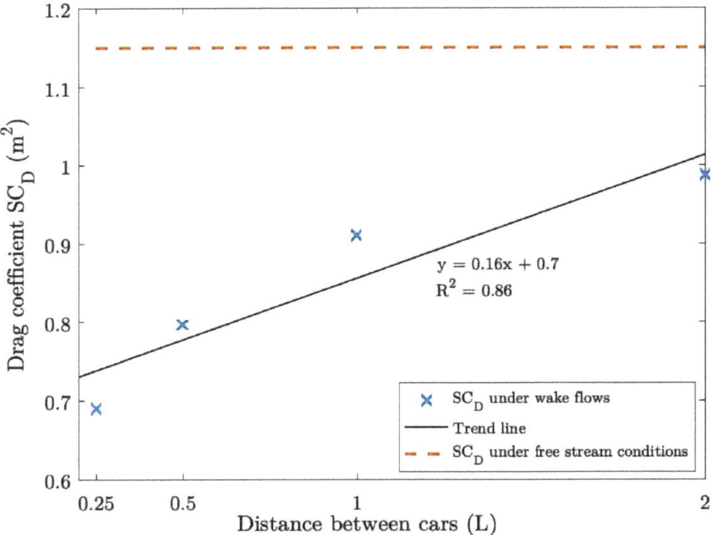

Figure 13. Comparison between values of SC_D under wake flows and under free stream conditions.

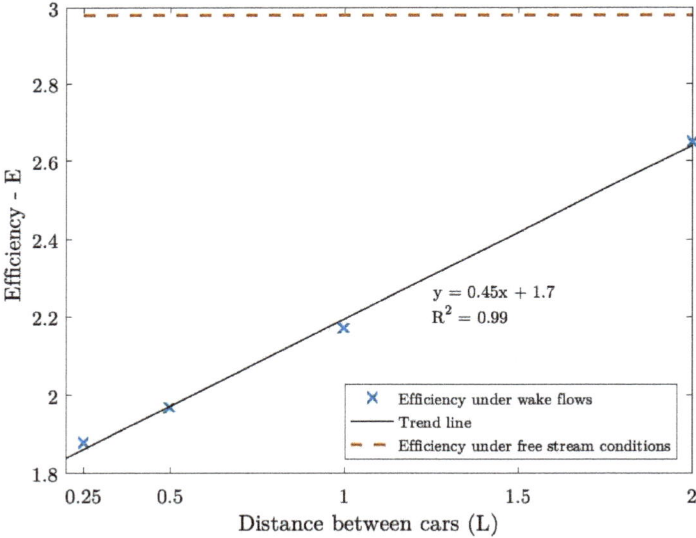

Figure 14. Comparison between values of efficiency under wake flows and under free stream conditions.

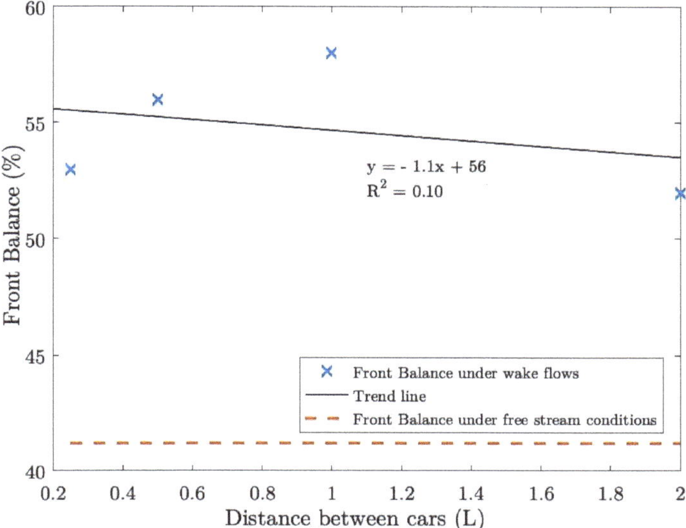

Figure 15. Comparison between values of front balance under wake flows and under free stream conditions.

Table 5. Percentage of change of the aerodynamic coefficients of the second car as regards the first car.

Distance	SC_L (m^2)	SC_D (m^2)	$\lvert C_L/C_D \rvert$	FB
0.25 L	−62%	−40%	−37%	+28.5%
0.5 L	−54.1%	−30.7%	−34%	+35.8%
1 L	−42.3%	−21%	−27%	+40.7%
2 L	−23.5%	−14.2%	−11%	+26.1%

The obtained results show that the reduction in the aerodynamic coefficients is clearly visible from an initial distance of two car lengths (approximately 10.6 m) to the closest case studied of 0.25 L (less than 1.5 m). The reduction of downforce ranges from a −23.5% to a very significant −62% in the worst case scenario. In a similar progression, the drag is reduced from a −14.2% to a −40% and, for this reason, so does the overall efficiency of the second car.

Besides the distinguished loss of downforce, the second car experiences a dramatic increase of front balance (FB) from +26% to 40%. This sudden increase on the front aerodynamic loads may presumably lead to experiencing oversteer (oversteer is caused when a car steers more than intended, thus losing the rear end) and safety issues while braking and on high-speed corners [28].

In general traits, it can be seen that, as the second car approaches and gets closer to the leading one, the aerodynamic loads are reduced, which worsens the performance of the car, but so does the drag. This is energetically a key point, as the power that is required to overcome the drag is smaller when the distance between the cars gets closer, which enables less fuel consumption for the car behind. Table 6 shows a hypothetical situation on the main straight of "Circuit de Barcelona", Catalunya, where the power and the energy that are required to overcome the different situations are evaluated (it has been assumed a distance of 1 km and a car speed of 50 m/s along the whole straight).

These same conclusions can be easily extrapolated to current road cars, although the conditions and data may differ notably, but not the overall conclusions.

Table 6. Energetic parameters required to overcome the air resistance.

Distance	Power Required (kW)	Energy Required (kJ)
0.25 L	33.44	668.8
0.5 L	40.39	807.84
1 L	50.77	1015.52
2 L	67.32	1346.4
Free stream	88	1760

Figure 16 is aimed at showing the obtained data in a visual representation, so that is possible to appreciate the rate of change in the various studied parameters.

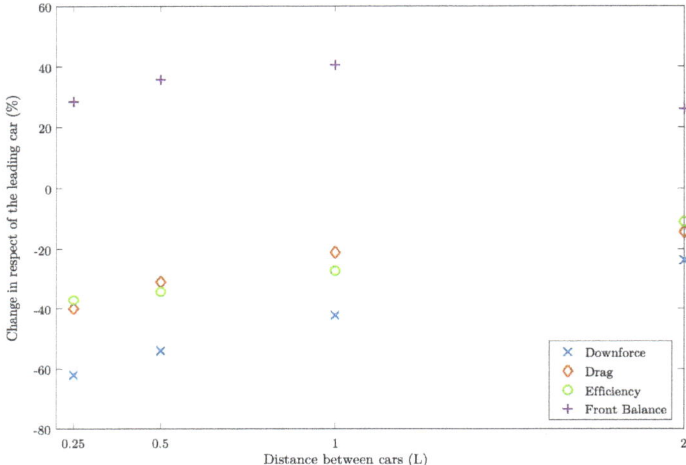

Figure 16. Percentage change of the performance of the second car in respect of the leading car.

As mentioned, the increase of the front balance levels on the second car enables a short, but rather interesting discussion. It is known that the weight distribution of a F1 2017 specification car is around 45.5% on the front axle [29], so the car is not supposed to exceed this 45.5% of front balance, as it may lead to stability concerns. The Center of Pressure, which, by definition, is such where the total sum of pressure fields act on, should always remain behind the Center of Gravity. This can be explained, as the yawing moment of the aerodynamic forces counterbalances the steering of the driver and, therefore, stabilises the car.

On the other hand, if the Center of Pressure is ahead of the Center of Gravity, then the yawing moment increases the sideslip angle and produces instability. As the obtained results show, the front balance of the second car adopts always values that are greater than 45.5% (see Table 5). This implies not only a reduction of the stability and the performance of the car (slower laptimes and more degradation of the front tires), but also a more challenging approach when driving the car; major ease of spinning and safety issues.

Moreover, the study analyses the performance of the most relevant aerodynamic devices on the follower car: front wing, rear wing, and diffuser.

Table 7 reflects that the loss of downforce of the front wing starts to appear severely at a distance of 1 L up until a very critical −38%, when it reaches 0.25 L. On the other hand, the drag levels are found to be grater than the leading car at a distance of two car lengths, but a similar behaviour to the downforce is found as long as the distance is decreased. This could be explained due to the turbulent conditions that are found far away from the leading car. Moreover, as the second car approaches the leading one, it enters into a very strong wake region that initially hits the front wing and eventually influences all behaviour of the car.

Table 7. Percentage of change of the aerodynamic coefficients in the front wing of the second car as regards the first car.

Distance Front Wing	SC_L (m^2)	SC_D (m^2)
0.25 L	−38%	−36.1%
0.5 L	−29%	−19.3%
1 L	−10.7%	−1.1%
2 L	−3.6%	+3%

This is precisely seen in Figure 17, where it is possible to see the behaviour commented earlier, as the low pressure zones of the front wing suffer an abrupt change and loss as the car enters into the closer wake region (cases 0.25 L and 0.5 L, where the tendency of the pressure distribution appears slightly inconsistent, although the numerical results do not). It is not at all elementary to try to find a reason why it occurs, but the most sensible explanation lies in the chaotic and unpredictable nature of turbulence. Besides, the larger pressure area located in 0.5 L is found in the very first tip of the front wing, which corresponds to the first element that is encountered by the airflow. In general, these changes are noticeably far less progressive than the rear wing's, which also evidences the premature front balance shift, but allows a more robust behaviour at greater distances.

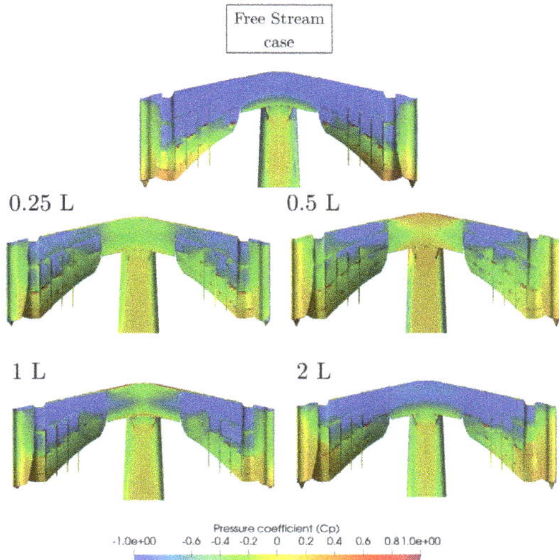

Figure 17. Pressure coefficient distribution on the front wing of the second car at a distance of 0.25 L, 0.5 L, 1 L, and 2 L as compared to the free stream case.

As for the streamlines of the velocity, Figure 18 (2 L) shows that the magnitude of the velocity is somehow similar to the one of the free stream flow, although it presents low velocity areas on the central part of the wing. However, endplates and wing tip elements are still useful for redirecting the flow towards the rear. As the distance is halved, the velocity of the flow is reduced and the front wing loses its capacity to govern the airflow, but it is not until cases 0.5 L and 0.25 L that the front wing is fed by really low kinetic energy flow that leaves it notably inoperative. The effectiveness of the generation of vortices and redirection of the flow is insignificant, just as checked in Table 7 due to the strong wake region.

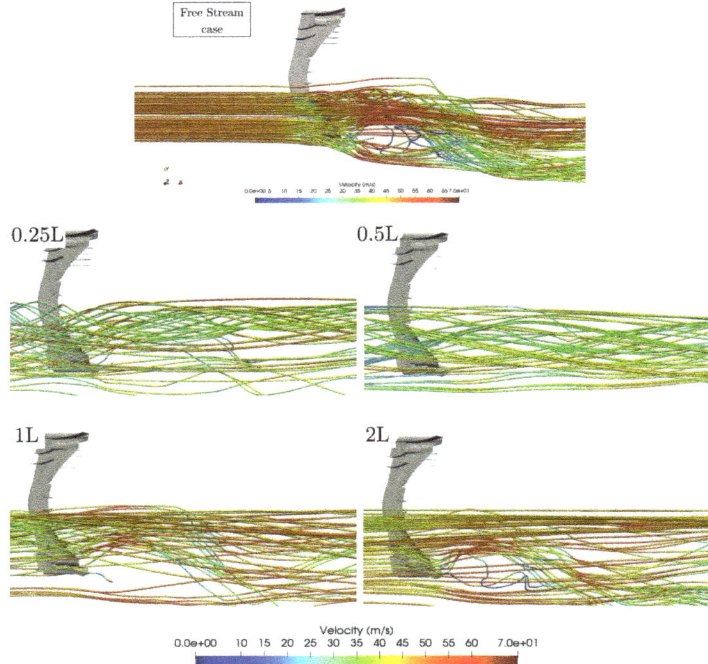

Figure 18. Streamlines of the velocity. Front wing of the second car at a distance of 0.25 L, 0.5 L, 1 L, and 2 L as compared to the free stream case.

Shifting to the rear wing, Table 8 portrays that the behaviour of the rear wing is notably different: the loss of downforce is very notable since the very first distance of 2 L and it keeps increasing as the slipstream distance gets closer. Similarly, the drag levels are also strongly reduced from the beginning and matching a comparable ratio that keeps the efficiency almost constant.

It can be stated then, that the rear wing is more affected under the wake effects than the front wing, as the latter seems to suffer less and only under close proximity. This somehow explains the increase of front balance (and forward change of the center of pressure) and its posterior decrease, as sketched in Figure 16.

Table 8. Percentage of change of the aerodynamic coefficients in the rear wing of the second car as regards the first car.

Distance Rear Wing	SC_L (m^2)	SC_D (m^2)
0.25 L	−57.9%	−53.3%
0.5 L	−54.8%	−50.6%
1 L	−52%	−48%
2 L	−40.3%	−36.2%

On the pressure side, from an early stage, one can see that the pressure loss in the rear wing is severely and gradually appreciated. The approach to the leading car distinctly damages the functioning of both its stabilising and suction purposes. Figure 19 sketches these comments.

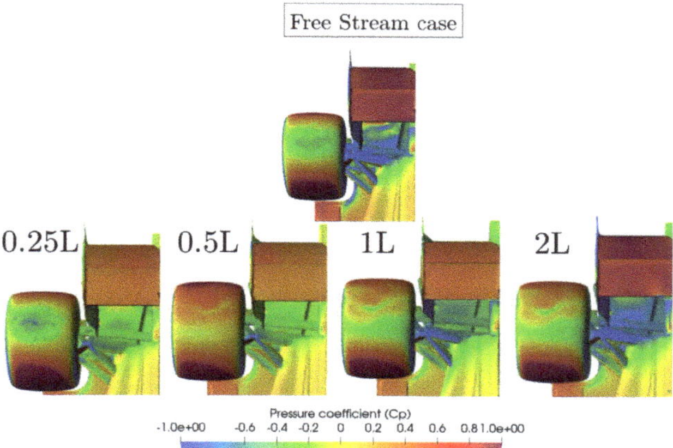

Figure 19. Pressure coefficient distribution on the rear wing of the second car at a distance of 0.25 L, 0.5 L, 1 L, and 2 L as compared to the free stream case.

Similarly, under a normal regime, Figure 20 shows that the rear wing works perfectly with aligned flow that is redirected by other aerodynamic devices, but it is possible to see that, as the distance is reduced, the streamlines tend to deflect slightly inboard. The clear jump in terms of velocity magnitude is produced from case 1 L to 0.5 L, where the flow experiences a moderate deceleration as it enters into the strongest wake region. It could be speculated that the rear wing is more sensitive to the direction of the flow than the front wing due to its low aspect ratio, the pronounced curvature, and the massive endplates at both sides. Figure 20 (0.25 L) shows a mixture of velocity flows with very low speed that unexpectedly create an inwash phenomena at the rear end of the wing. This evidences that the whole aerodynamic package of the follower car is prominently disrupted under the wake conditions.

Finally, Table 9 reports that the diffuser is the device that suffers the most under the wake conditions, as the downforce loss that is in close proximity is around 70% and the drag is found to be reduced by around 57% at the same distance. However, the interesting conclusion is that the reduction of downforce starts from the very first beginning and keeps increasing as the distance is reduced.

This evidences that overtaking maneuvers are heavily influenced since the start (it is important not to forget that the diffuser is the greatest source of downforce generation, as seen in Table 4).

This generates an interesting and deep discussion, as it is encountered that the greatest way of generating aerodynamic loads under a free stream flow, is, at the same time, the worst one under wake flows. The whole conception and functioning of the underbody is found to be absolutely pointless; therefore, other aerodynamic paths and solutions should be evaluated if these losses are wished to be recovered.

On the pressure side, the results that are presented in Figure 21 again show that the whole low pressure zone is affected since the very first beginning, with moderate areas near the diffuser strakes with higher pressure values. However, the degradation of the performance is perfectly noticeable and again proves that the diffuser suffers excessively under wake flows until its contribution becomes almost negligible.

As for the streamlines of the velocity, Figure 22 describes how the underbody, and particularly the diffuser, is affected by the wake flow. In free stream conditions, the diffuser is fed by a high energised flow that is redirected by the front wing and guided around the flat floor. Nonetheless, the streamlines of the flow are not completely straight, as the vortices management allow for the control of the airflow around the underbody. Figure 22 (2 L) displays a quite non-disturbed behaviour of the airflow, as the management of it is still acceptable. When the distance is reduced, the performance of the underbody starts worsening due to the flow arriving more disturbed into the diffuser, hence not being able to operate properly. Smaller distances, such as cases 0.5 L and 0.25 L, reflect an underbody region that is fed by a very low kinetic and rotational flow.

The main reason for the massive downforce losses that are reported in Table 9 is that the underfloor is notably sensitive under wake flows, as it works closer to the ground than the front wing. This means that the low energy (and highly-rotational) airflow may not be compressed around this small area, therefore experiencing a massive performance loss.

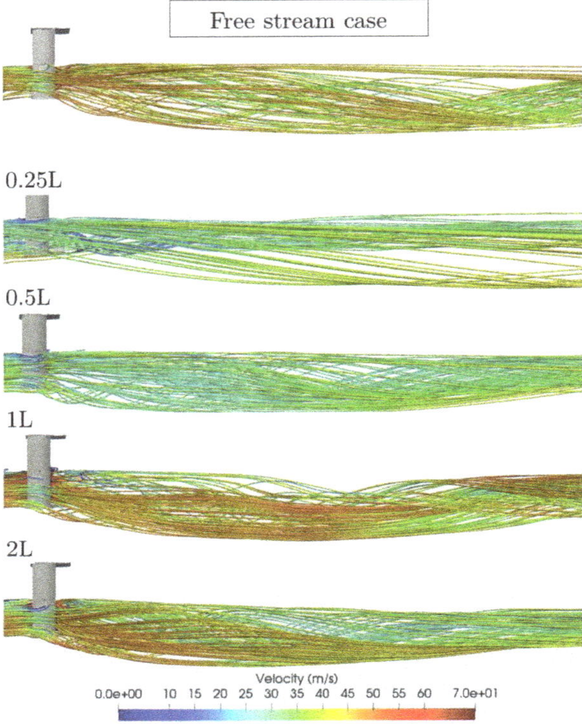

Figure 20. Streamlines of the velocity. Rear wing of the second car at a distance of 0.25 L, 0.5 L, 1 L, and 2 L as compared to the free stream case.

Table 9. Percentage of change of the aerodynamic coefficients in the diffuser of the second car as regards the first car.

Distance Diffuser	$SC_L\ (m^2)$	$SC_D\ (m^2)$
0.25 L	−70.2%	−57.2%
0.5 L	−62.8%	−48.7%
1 L	−46.1%	−29%
2 L	−25.3%	−16.9%

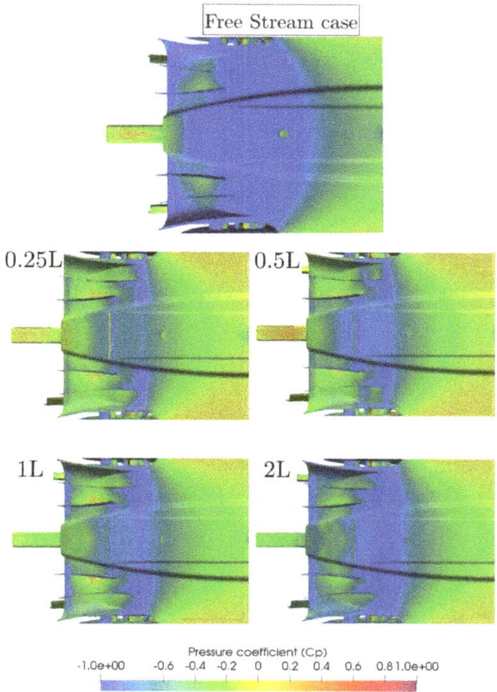

Figure 21. Pressure coefficient distribution on the diffuser of the second car at a distance of 0.25 L, 0.5 L, 1 L, and 2 L as compared to the free stream case.

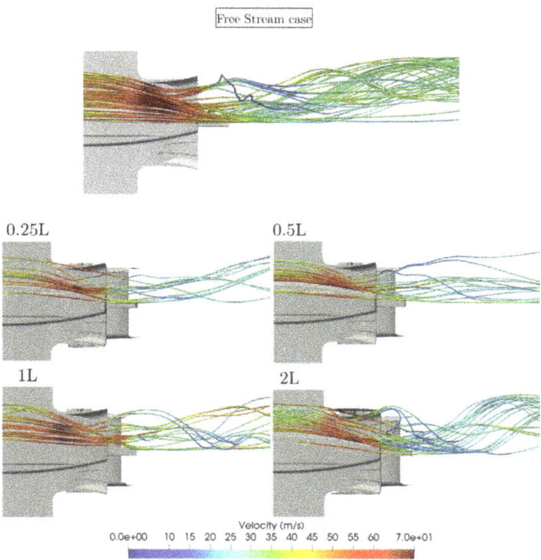

Figure 22. Streamlines of the velocity. Diffuser of the second car at a distance of 0.25 L, 0.5 L, 1 L, and 2 L as compared to the free stream case.

Finally, Figure 23 shows the plots of Tables 7–9 in order to appreciate the aforementioned changes in the aerodynamic coefficients of the different elements.

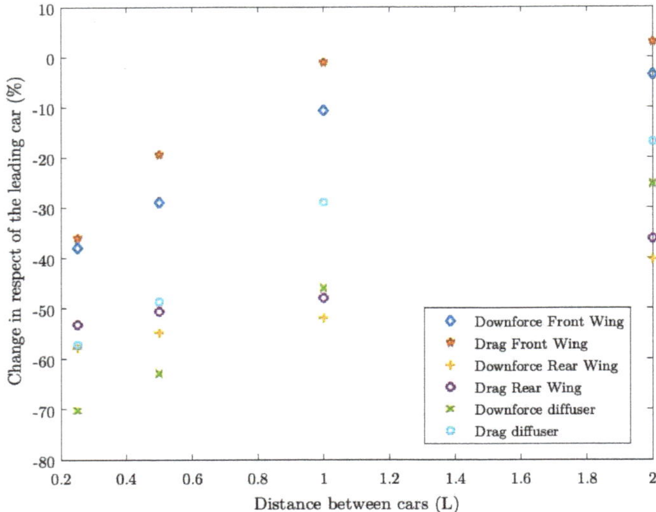

Figure 23. Percentage of change on the performance of the second car.

On the wake side, these effects can easily be appreciated in Figure 24, where the second car is affected by a flow that is lower in terms of kinetic energy, as the wake that is generated by the leading car is released far away disturbing its follower. It is also clear that, as the second car gets closer, it inherently enters into a unique wake structure characterised by very low speed flow, that ranges from 0 to 10 m/s, therefore, resulting severely affected. It is seen that as the second car reduces the distance, its wake originates a separation region that enlarges and becomes evident as the distance is closed. At a large distance (2 L), the second car's wake adopts a needle shape, which somehow imitates the free stream natural wake, but this shape soon disappears at closer distances.

As for the leading car, it can be noted that its wake is not notably modified (in terms of shape and contours) by the presence of the follower car. The aerodynamic coefficients that are evaluated on the leading car only experience a low variation —less than 3%—when the distance between the two cars is set at 0.25 L.

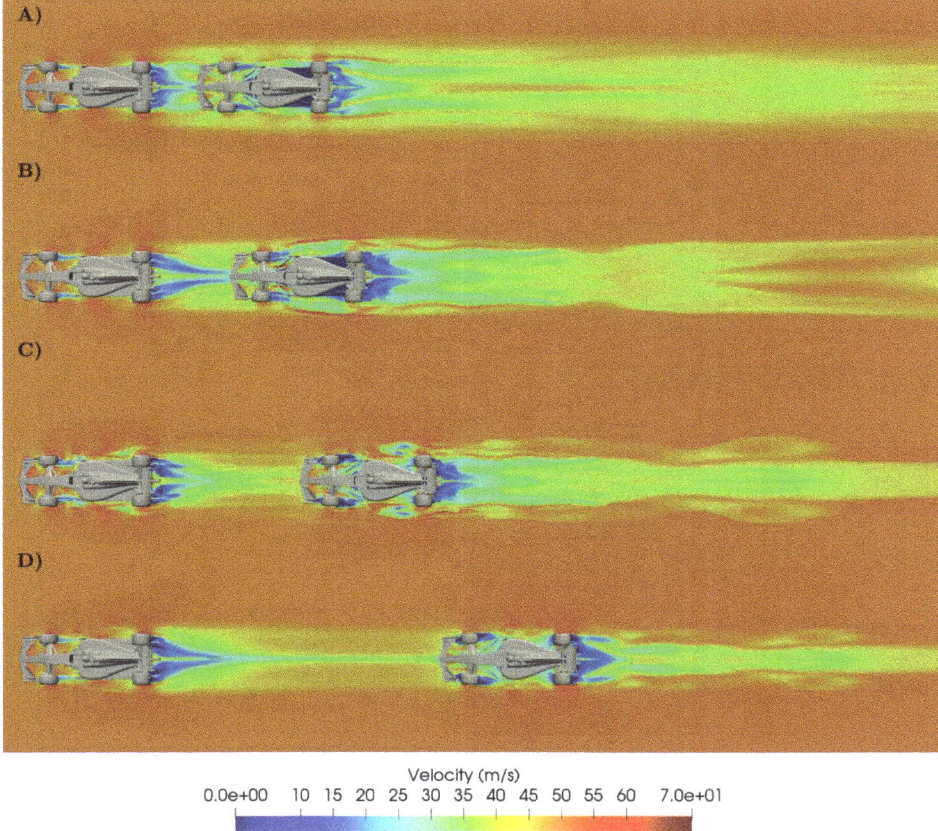

Figure 24. Velocity contours on a top plane at a distance of (**A**) 0.25 L, (**B**) 0.5 L, (**C**) 1 L, and (**D**) 2 L.

Shifting now to the streamlines of the velocity, Figure 25 displays two different planes for each scenario. The first thing that one may notice is how the airflow is perfectly attached to the first car; from the front wing, going through the sidepods and bodywork and finally exiting the rear wing.

However, it is possible to also see that the exiting airflow on the rear-end of the leading car is somehow divided into two characteristic flows: the first one, located on the superior area, which adopts high-speed values due to being accelerated around the bodywork (low pressure zone and smooth behaviour), and the second one, which exits the diffuser upwards and is mainly a turbulent flow continuously undergoing changes in both magnitude and direction, as reported in [30]. The mixture of the previously mentioned flows is what originates such a chaotic wake region, as it is formed by the combination of multiple flows with various natures and velocities [31].

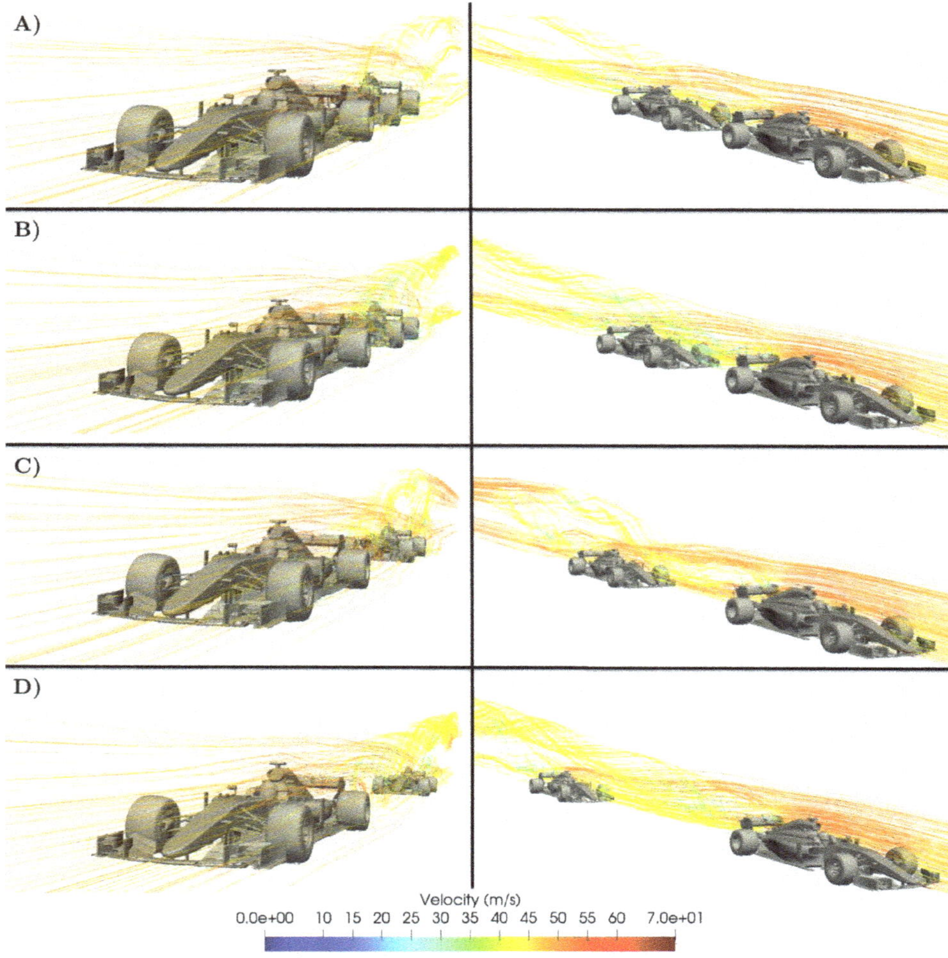

Figure 25. Streamlines of the velocity at a distance of (**A**) 0.25 L, (**B**) 0.5 L, (**C**) 1 L, and (**D**) 2 L.

4. Conclusions

The initial purpose of the present study was to establish whether the current F1 regulations required an urgent aerodynamic change in order to objectively improve the category. In other words, to test and validate, by means of CFD analyses, that the upcoming regulation changes are aerodynamically justified. Accordingly, the results gathered in the present manuscript may be useful to F1 teams, FIA administrations, and to anyone interested in racing aerodynamics.

The obtained results suggest that modern F1 cars are designed and well optimised to run under free stream flows (as the reference data coincides with the numerical results), but they suffer excessively when running under wake flows. Overall, the aerodynamic loads tend to be reduced when running under close proximity, ranging from 23% to a very significant 62% in the closest case.

In regards to the individual focus that is placed on the different aerodynamic devices, it has been found that the front wing experiences a sudden jump on downforce losses only when it enters into the closer wake region. On the other hand, the rear wing massively suffers from long distances, but its losses are way more linear and moderate. As for the diffuser, it is found that it represents the most affected aerodynamic device, as its performance is reduced from a considerable 25% to a huge 70%. This evidences that the conception of the diffuser and vortex management under the floor becomes critically compromised under wake flow conditions.

The modern performance of Ground Effect by means of vortices management represents a very unique and complex way of modelling modern aerodynamics, but, at the same time notably compromise the performance of the cars when an overtaking maneuver is intended. For this reason, it is possible to guarantee that the FIA changes in the current regulations is considered to be adequate, necessary, and justified in terms of aerodynamic necessities.

However, it is essential to comment that the present study presents some research limitations, as CFD methodology always requires an experimental validation. Accordingly, directions in further research point towards experimental validation (involving the use of wind tunnels with movable ground and spinning tires) in order to ensure that the data gathered reflect the reality properly.

Author Contributions: A.G. configured and performed the numerical simulations as well as the analysis of the results and the paper elaboration. R.C. supervised the study and reviewed the final version of the paper. All authors have read and agreed to the published version of the manuscript.

Funding: This research received no external funding.

Conflicts of Interest: The authors declare no conflict of interest.

Abbreviations

The following abbreviations are used in this manuscript:

CFD	Computational Fluid Dynamics
FVM	Finite Volume Method
GAMG	Geometric Algebraic Multi Grid
GCI	Grid Convergence Index
FIA	Federation Internationale de l'Automobile
CAD	Computer Aided Design
RAM	Random Access Memory
L	Vehicle length [m]
ϕ	Reference parameter [m^2]
l	Turbulent length scale [m]
I	Turbulent intensity [%]
Re	Reynolds Number
Cp	Pressure coefficient
ρ	Fluid density [kg/m^3]
U_∞	Fluid velocity [m/s]
g	Gravitational acceleration [m/s^2]
μ	Dynamic viscosity [kg/m·s]
S	Reference Surface [m^2]
C_L	Lift coefficient
C_D	Drag coefficient
E	Aerodynamic Efficiency
FB	Front Balance

References

1. Wilson, M.R.; Dominy, R.G.; Straker, A. The aerodynamic characteristics of a race car wing operating in a wake. *SAE Int. J. Passeng. Cars-Mech. Syst.* **2008**, *1*, 552–559. [CrossRef]
2. Ravelli, U.; Savini, M. Aerodynamic Simulation of a 2017 F1 Car with Open-Source CFD Code. *J. Traffic Transp. Eng.* **2018**, *6*. [CrossRef]
3. Ravelli, U.; Savini, M. Aerodynamic investigation of blunt and slender bodies in ground effect using OpenFOAM. *Int. J. Aerodyn.* **2019**. [CrossRef]

4. Newbon, J.; Sims-Williams, D.; Dominy, R. Aerodynamic analysis of Grand Prix cars operating in wake flows. *SAE Int. J. Passeng. Cars. Mech. Syst.* **2017**, *10*, 318–329. [CrossRef]

5. Perry, R.; Marshall, D. An Evaluation of Proposed Formula 1 Aerodynamic Regulations Changes Using Computational Fluid Dynamics. In Proceedings of the 26th AIAA Applied Aerodynamics Conference, Honolulu, Hawaii, 18–21 August 2008; p. 6. [CrossRef]

6. Ogawa, A.; Yano, S.; Mashio, S.; Takiguchi, T.; Nakamura, S.; Shingai, M. *Development Methodologies for Formula One Aerodynamics*; Honda R&D Technical Review 2009; F1 Special (The third Era Activities): Sakura/Chiba, Japan, 2009.

7. Senger, S.; Bhardwaj, S.R. Aerodynamic Design of F1 and Normal Cars and Their Effect on Performance. *Int. Rev. Appl. Eng. Res.* **2014**, 2248–9967.

8. Larsson, T. 2009 Formula One Aerodynamics BMW Sauber F1.09—Fundamentally Different. In Proceedings of the 4th European Automotive Simulation Conference, Munich, Germany, 6–7 July 2009; p. 4.

9. Reddit. Ferrari SF70H, the 2017 Championship Contender. Available online: https://www.reddit.com/r/formula1/comments/bw9v7z/the_ferrari_sf70h_shark_fin_looks_good/ (accessed on 3 May 2020).

10. Newbon, J.; Dominy, R.; Sims-Williams, D. Investigation into the Effect of the Wake from a Generic Formula One Car on a Downstream Vehicle. In *Proceedings of the International Vehicle Aerodynamics Conference*; Elsevier Science: Amsterdam, The Netherlands, 2014.

11. Hajdukiewicz, M.; Geron, M.; Keane, M.M. Formal calibration methodology for CFD models of naturally ventilated indoor environments. *Build. Environ.* **2013**, *59*, 290–302. [CrossRef]

12. Williams, J.; Quinlan, W.; Hackett, J.; Thompson, S.; Marinaccio, T.; Robertson, A. A calibration study of CFD for automotive shapes and CD. *SAE Trans.* **1994**, *103*, 308–327.

13. Onshape a PTC Business. Perrinn Limited I Customer. Available online: https://www.onshape.com/customers/perrinn-limited (accessed on 3 May 2020).

14. OpenFOAM Foundation. OpenFOAM Resources I Documentation I OpenFOAM. Available online: https://openfoam.org/resources/ (accessed on 2 February 2020).

15. Constantin, P.; Foias, C. *Navier-Stokes Equations*; University of Chicago Press: Chicago, IL, USA, 1988.

16. Eymard, R.; Gallouët, T.; Herbin, R. Finite volume methods. *Handb. Numer. Anal.* **2000**, *7*, 713–1018. [CrossRef]

17. Menter, F.R.; Kuntz, M.; Langtry, R. Ten years of industrial experience with the SST turbulence model. *Turbul. Heat Mass Transf.* **2003**, *4*, 625–632.

18. Broniszewski, J.; Piechna, J. A fully coupled analysis of unsteady aerodynamics impact on vehicle dynamics during braking. *Eng. Appl. Comput. Fluid Mech.* **2019**, *13*, 623–641. [CrossRef]

19. Moukalled, F.; Mangani, L.; Darwish, M. *The Finite Volume Method in Computational Fluid Dynamics*; Springer: Berlin/Heidelberg, Germany, 2016; Volume 6.

20. Celik, I.B.; Ghia, U.; Roache, P.J.; Freitas, C.J.; Coleman, H.; Raad, P.E. Procedure for estimation and reporting of uncertainty due to discretization in CFD applications. *J. Fluids Eng. Trans. ASME* **2008**, *130*, 0780011–0780014. [CrossRef]

21. CFD Online. Turbulence Length Scale-CFD-Wiki, the Free CFD Reference. Available online: https://www.cfd-online.com/Wiki/Turbulence_length_scale (accessed on 1 May 2020).

22. CFD Online. Turbulence Intensity-CFD-Wiki, the Free CFD Reference. Available online: https://www.cfd-online.com/Wiki/Turbulence_intensity (accessed on 27 April 2020).

23. PERRINN CFD. Available online: https://docs.google.com/spreadsheets/d/1POHSdkdfXeUEeGgh-meufNzVRUycCLvzUfn_TVXs_PE/edit#gid=0 (accessed on 4 March 2020).

24. Soso, M.; Wilson, P. Aerodynamics of a wing in ground effect in generic racing car wake flows. *Proc. Inst. Mech. Eng. Part D J. Automob. Eng.* **2006**, *220*, 1–13. [CrossRef]

25. Vargas, R.V. Meteorología Sinóptica—Vorticidad. Available online: https://slideplayer.es/slide/13178976/ (accessed on 9 May 2020).

26. Nakagawa, M.; Kallweit, S.; Michaux, F.; Hojo, T. Typical velocity fields and vortical structures around a formula one car, based on experimental investigations using particle image velocimetry. *SAE Int. J. Passeng. Cars-Mech. Syst.* **2016**, *9*, 754–771. [CrossRef]

27. Ehirim, O.; Knowles, K.; Saddington, A. A Review of Ground-Effect Diffuser Aerodynamics. *J. Fluids Eng.* **2018**, *141*, 020801. [CrossRef]

28. Howell, J. Catastrophic lift forces on racing cars. *J. Wind. Eng. Ind. Aerodyn.* **1981**, *9*, 145–154. [CrossRef]

29. McBeath, S. *Competition Car Aerodynamics*, 3rd ed.; Veloce Publishing: Dorchester/Poundbury, UK, 2017.
30. Dominy, R. The influence of slipstreaming on the performance of a Grand Prix racing car. *Proc. Inst. Mech. Eng. Part D J. Automob. Eng.* **1990**, *204*, 35–40. [CrossRef]
31. Bleacher Report, Inc. F1 Turbulence: Dumbing It Down | Bleacher Report |. Available online: https://bleacherreport.com/articles/220724-f1-turbulence-dumbing-it-down (accessed on 4 April 2020).

 *energies*

Article

Study of Surface Roughness Effect on a Bluff Body—The Formation of Asymmetric Separation Bubbles

Alex Mendonça Bimbato [1],* **, Luiz Antonio Alcântara Pereira [2]** **and Miguel Hiroo Hirata [3]**

[1]   School of Engineering, São Paulo State University (UNESP), Guaratinguetá SP 12.516-410, Brazil
[2]   Mechanical Engineering Institute, Federal University of Itajubá (UNIFEI), Itajubá MG 37.500-903, Brazil; luizantp@unifei.edu.br
[3]   Faculty of Technology, State University of Rio de Janeiro (FAT-UERJ), Resende RJ 27.537-000, Brazil; miguelhhirata@gmail.com
*   Correspondence: alex.bimbato@unesp.br; Tel.: +55-12-3123-2874

Received: 31 October 2020; Accepted: 18 November 2020; Published: 20 November 2020

Abstract: Turbulent flows around bluff bodies are present in a large number of aeronautical, civil, mechanical, naval and oceanic engineering problems and still need comprehension. This paper provides a detailed investigation of turbulent boundary layer flows past a bluff body. The flows are disturbed by superficial roughness effect, one of the most influencing parameters present in engineering applications. A roughness model, recently developed by the authors, is here employed in order to capture the main features of these complex flows. Starting from subcritical Reynolds number simulations (Re = 1.0×10^5), typical phenomena found on critical and supercritical flow regimes are successfully captured, like non-zero lift force and its direction change, drag crisis followed by a gradual increase on this force, and separation and stagnation points displacement. The main contribution of this paper is to present a wide discussion related with the temporal history of aerodynamic loads of a single rough circular cylinder capturing the occurrence of asymmetric separation bubbles generation. The formation of asymmetric separation bubbles is an intrinsic phenomenon of the physical problem, which is successfully reported by our work. Unfortunately, there is a lack of numerical results available in the literature discussing the problem, which has also motivated the present paper. Previous study of our research group has only discussed the drag crisis, without to investigate its gradual increase and the change on lift force direction. Our results again confirm that the Lagrangian vortex method in association with Large-Eddy Simulation (LES) theory enables the development of two-dimensional roughness models.

Keywords: bluff body aerodynamics; roughness model; boundary layer separation; vortex shedding; Lagrangian vortex method

---

1. Introduction

The technologic and science enhancement in many fields of knowledge is because of deep analysis and consequent comprehension of flow fundamentals around bodies of arbitrary shape. Many of the advances observed in ocean, civil and wind engineering have been possible because of studies conducted over time in the field of aerodynamics. The importance of knowing and mastering this subject can be easily recognized when it is observed that, most of bodies present in situations of practical interest for engineering, are exposed to air or water flows. These flows are typical examples of fluid-structure interaction problems, where the transition to turbulence is undeniably a complex phenomenon of nonlinear hydrodynamics. Such phenomenon arouses great scientific interest, has a great impact on engineering applications (aeronautical, civil, mechanical, naval and oceanic), and has still its origin linked to the development of instabilities associated with the interaction of two shear layers of opposite signals; the latter are caused by separation points originated from the bluff body surface [1], like a circular cylinder, which is commonly studied because of its rich collection of flow phenomena. On the other hand, the transition to turbulence originates in the viscous wake region and, as the Reynolds number, Re, increases, it occurs more and more close to the body surface, passing through the two shear layers formed from the separation points (commonly defined by an angle θ_{sep}), until reaching the boundary layer region of the body.

In the literature, there are traditional works discussing transition to turbulence e.g., [2–10], where the surface roughness influence has also been included. The current scope of the research theme about "surface roughness effect" indicates few experimental and numerical studies e.g., [11–18]. The importance of the more relevant works will be discussed throughout the paper. The occurrence of the transition to turbulence into the boundary layer region is particularly important for practical applications, since the flow has a greater amount of momentum supporting the adverse pressure gradient; furthermore, the separation points move downstream of the body surface, causing the so-called drag crisis, which consists of a reduction in the drag force, that is opposed to body motion. The physics involved in this whole process is so complex, and under certain Reynolds number, the separation of the boundary layer is followed by a reattachment and a new separation again; the closed thin separated region was termed in the classical literature as a separation bubble.

According to Tani [2], the formation of separation bubbles causes low drag force values. Although the circular cylinder is a symmetrical body, the formation of separation bubbles does not occur simultaneously on the upper and lower surfaces of the body, which causes the appearance of an asymmetric pressure distribution and, as consequence, a non-zero lift force, contrary to what is expected to be obtained in any flow around a symmetrical bluff body.

The physical phenomenon above mentioned manifests at high Reynolds numbers, so that the presence of small disturbances in the flow will be amplified changing the bluff body aerodynamics. One of the types of disturbance most encountered in practical engineering problems refers to the surface roughness effect of a body, which affects heat exchangers efficiency, ship propellers performance, aerodynamic of sport materials, flow around offshore structures and wind turbines performance. The use of surface roughness effect on fluid flow requires to change the body geometry, being, therefore, classified as a passive method of vortex shedding control.

In Achenbach [3], the surface roughness effect is experimentally studied on the flow past a circular cylinder. He used a high-pressure wind tunnel, where high Reynolds number flows up to $Re = 3 \times 10^6$ could be obtained. It was presented the classical $C_D \times Re$ curve (C_D is the drag coefficient) for the flow past a circular cylinder dividing it into four flow regimes: subcritical ($Re < 2.0 \times 10^5 - 5.0 \times 10^5$), critical ($Re \cong 2.0 \times 10^5 - 5.0 \times 10^5$), supercritical ($2.0 \times 10^5 - 5.0 \times 10^5 < Re \leq 3.5 \times 10^6$) and transcritical ($Re > 3.5 \times 10^6$); the latter is nowadays called pos-critical, as will be illustrated latter. According to him, the surface roughness effect does not affect the subcritical flow regime. On the other hand, as the Reynolds number increases, the drag force suddenly drops until reach a minimum value; this range of Reynolds number is within the critical flow regime. Further, when the Reynolds number exceeds the minimum value of the drag force, the drag force grows up again (supercritical flow regime) reaching a nearly constant value (transcritical flow regime). It was also observed that the drag force, within the transcritical flow regime, becomes bigger with increasing roughness. In regarding of the percentage of the skin friction (or viscous) component of the drag force, it was about 1% in the subcritical flow regime and 3% for the other three flow regimes.

Bearman [5] examined the flow past a smooth circular cylinder for the Reynolds number ranging from $Re = 1.0 \times 10^5$ to 7.5×10^5 aiming to investigate the base pressure coefficient behavior depending on the Reynolds number. An interesting feature identified by him is that drag coefficient has almost the same value of the base absolute pressure value. Besides, according to the experimental data, the separation bubble formed on one side of the body provoked a discontinuity recorded at

Re = 3.4×10^5 of the base pressure coefficient; during this phase, the lift coefficient, C_L, calculated from the pressure distribution, was about 1.3. The separation bubble on the other side formed at Reynolds number flow of Re = 3.8×10^5, and the lift coefficient was C_L=0.0, being measured as soon as this happens and remains in this way, as the Reynolds number increases.

Guven et al. [7] presented in one same diagram the results obtained in their experimental tests with other results available in the literature [3,8,9] for circular cylinders with different roughness heights. It was showed scatter results with up to 60%, even for the same roughness and Reynolds numbers. This happens because, although the Reynolds number governs this kind of flow, it is so difficulty to guarantee that two experimental tests have exactly the same roughness surface; besides that, there are some influencing parameters which can cause a premature initiation of a flow regime or even its modification.

According to Zdravkovich [10], the above referred influencing parameters are: (a) free stream turbulence, described by its intensity and scale; (b) surface roughness effect, described by its mean relative surface roughness, ε^*/D^*, and its texture; (c) wall confinement (blockage effect); (d) aspect ratio; (e) presence of other boundary near the body; (f) end condition of the body; (g) body oscillation effect. The free stream turbulence and surface roughness influence are the most common disturbances in all engineering applications. The wall confinement and the aspect ratio are important influencing parameters that interfere in the comparison between experimental and numerical results, especially when the surface roughness effect is an influencing parameter too [17,18]. These features help us to understand the scatter in the experimental data presented by Guven et al. [7].

In a recent paper, Bimbato et al. [14] developed a Lagrangian vortex method using two-dimensional roughness model, which was based on the physics involved in turbulent boundary layer flows. The roughness surface effect was simulated without make changes on geometry of the body, what means that the investigated circular cylinder surface remains smooth. The key idea is that a point set strategically located close to the body surface injects momentum in its boundary layer aiming to represent surface roughness effect. Their strategy successfully shows that the separation of boundary layer is delayed. They studied the incompressible flow around both smooth and rough circular cylinders at Re = 1.0×10^5, and it was only reported the drag crisis. Based on aerodynamic loads computations, it was concluded that their methodology, starting from a typical subcritical Reynolds number flow, was able to capture supercritical flow patterns. Overall, the drag coefficient computed just fall and it did not present the gradual increase, which is a typical feature of supercritical flows. Also, it was not presented results about the lift force, which would be interesting since a change on the lift force direction is expected [2].

The present paper, therefore, contributes to the literature using the same roughness model developed by Bimbato et al. [14] in order to capture physical details concerning the complex flow around a single rough circular cylinder. Put in other words, the main contribution here states on physical discussions involving the drag and lift forces connected with the occurrence of asymmetric separation bubbles generation. The generation of asymmetric separation bubbles is an intrinsic phenomenon to the physics of the problem [2]. In Section 4, it will be reported that our numerical results have captured that phenomenon, which has rarely been discussed by other numerical investigations. The results also confirm that two-dimensional Lagrangian vortex method with both LES and roughness models is able to capture not only the drag crisis, but also the non-zero lift force, which is typical of critical Reynolds number flows, and the gradual increase on drag coefficient, which is a typical feature of supercritical flows. Furthermore, our results show a delay in the boundary layer separation and also a stagnation point displacement.

Finally, it can be said that numerical simulations using Lagrangian vortex method have been employed to study problems in several areas of engineering. It is possible to mention a large number of studies, from the classics e.g., [19–22] to the most current ones e.g., [23,24]. Despite the high Reynolds numbers involved in practical engineering problems, the use of two-dimensional numerical simulations is necessary not only to develop and validate computational codes, but also to help understand the complex physical phenomena involved in such flows, and also to help conservative designs for engineering applications. A wide range of two-dimensional studies can be found in literature e.g., [14,17,18,25–31] which includes relevant contributions of our research group.

2. Mathematical Formulation

Figure 1 illustrates the two-dimensional fluid domain, Ω, to be studied, and defined by surface S=$S_1 \cup S_2$ (S_1 refers to the circular cylinder boundary and S_2 is the boundary far from the body). In the same figure, U* is the incompressible inlet flow, D* defines the outer cylinder diameter, θ denotes upper angular position, x* and y* are the global coordinate system, and the symbol * means dimensional quantities.

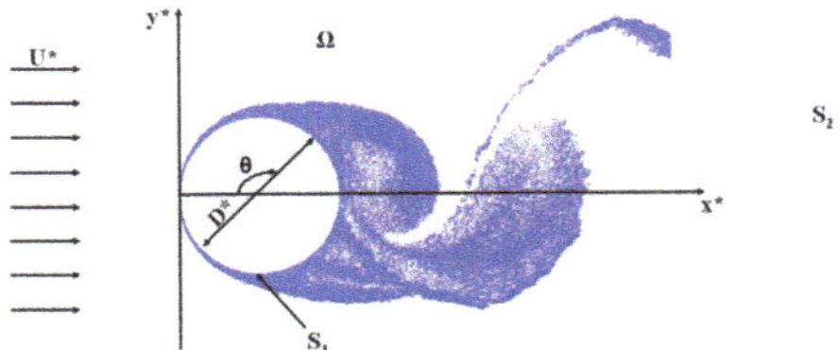

Figure 1. Definition of the fluid domain for the investigated problem, Ω.

The flow to be studied is assumed turbulent and, thus, a LES modeling is employed to disconnect the large eddies from the smaller ones. As practical result, the continuity and the momentum equations can be filtered resulting in Equations (1) and (2), such as, respectively:

$$\frac{\partial \overline{u_i^*}}{\partial x_i^*} = 0 \tag{1}$$

$$\frac{\partial \overline{u_i^*}}{\partial t^*} + \frac{\partial}{\partial x_j^*}\left(\overline{u_i^* u_j^*}\right) = -\frac{1}{\rho}\frac{\partial \overline{p^*}}{\partial x_i^*} + 2\frac{\partial}{\partial x_j^*}\left[(\upsilon + \upsilon_t)\overline{S}_{ij}^*\right] \tag{2}$$

In Equations (1) and (2), the *i*-th filtered velocity component of the flow field is identified by $\overline{u_i^*}$ (being $u_i^* = \overline{u_i^*} + u_i'^*$ with $u_i'^*$ representing the velocity fluctuation), ρ defines the density of fluid, $\overline{p^*}$ represents the pressure filtered field ($p^* = \overline{p^*} + p'^*$; p'^* is related to the pressure fluctuation), υ defines the fluid kinematic viscosity, υ_t defines the local eddy viscosity coefficient, and $\overline{S}_{ij}^*$ characterizes the deformation tensor of the filtered field [32].

The present methodology generates vortex blobs around the body surface during each time stepping aiming, thus, to solve the large eddy phenomena through the Equations (1) and (2). On the other hand, the small eddy phenomena are simulated using the concept of eddy viscosity coefficient (a typical LES approach). According to Lesieur and Métais [33], the local eddy viscosity coefficient at a point x* in the flow field, and at an instant t*, can be calculated via the local kinetic energy spectrum:

$$\upsilon_t\left(x^*,\ \Delta^{+*},\ t^*\right) = 0.105 C_k^{-3/2}\Delta^{+*}\sqrt{\overline{F}_2^*(x^*,\ \Delta^{+*},\ t^*)} \tag{3}$$

being $C_k = 1.4$ the constant of Kolmogorov and $\overline{F}_2^*(x^*,\ \Delta^{+*},\ t^*)$ the local second-order velocity structure function of the filtered field; the latter is defined as follows [33]:

$$\overline{F}_2^*\left(x^*,\ \Delta^{+*},\ t^*\right) = \left\|\overline{\vec{u}^*}(x^*,\ t^*) - \overline{\vec{u}^*}(x^* + r^*,\ t^*)\right\|^2_{|r^*|=\Delta^{+*}} \tag{4}$$

In Equation (4), and for the three dimensional space, $\overline{\vec{u}^*}(x^*,\ t^*) - \overline{\vec{u}^*}(x^* + r^*,\ t^*)$ represents the average speed differences between the center of a sphere of radius $|r^*| = \Delta^{+*}$, located at x*, and points placed on the sphere surface (Δ^{+*} separates the scales that are computed by the numerical method of the ones that should be modeled). In this approach the small scales are assumed to be homogeneous and isotropic, and the point of the flow field where one wants to quantify the turbulent activity is assumed as the center of the sphere [33].

The impermeability boundary condition on the cylinder surface (at S_1 in Figure 1) is established such as:

$$\vec{u}_n^* - v_n^* = 0 \text{ on } S_1 \tag{5}$$

being $\vec{u}_n^*$ the fluid normal velocity and v_n^* the normal velocity of the body surface.

The no-slip boundary condition on the cylinder surface (at S_1 in Figure 1) is imposed through the following equation:

$$\vec{u}_\tau^* - v_\tau^* = 0 \text{ on } S_1 \tag{6}$$

being $\vec{u}_\tau^*$ the fluid tangential velocity and v_τ^* the tangential velocity of the body surface.

The far away boundary condition is given by:

$$\left|\vec{u}\right| \rightarrow U^* \text{ on } S_2 \text{ (Figure 1)} \tag{7}$$

It is very practical to make dimensionless all the quantities previously presented, aiming to obtain generality gain from the numerical simulations. Thus, in the above equations, D^* and U^* are assumed as length and velocity scales, respectively. As consequence, the non-dimensional time is defined by $t = t^*U^*/D^*$.

3. The Lagrangian Vortex Method

In order to solve numerically the mathematical problem presented on Section 2, the first step is to discretize the circular cylinder surface shown in Figure 1. This is done through the so-called panel method [34], which consists on discretizing any solid boundary in panels, over which singularities are distributed in order to guarantee one of the boundary conditions presented in Section 2 at the "pivotal point" (center point) of each panel (Figure 2). In the present approach, the cylinder surface is discretized and represented by NP flat panels with source distribution of constant density. The sources generation is necessary to ensure the impermeability boundary condition Equation (5) in association with the mass conservation of the problem.

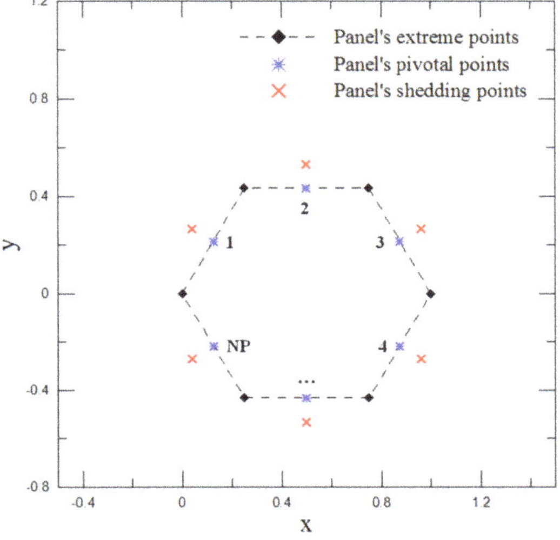

Figure 2. Circular cylinder surface discretized by NP flat panels (in this example, NP = 6).

The Lagrangian vortex method is nothing more than a boundary layer model, which consists on discretizing a flow property (in this case, the vorticity filtered field) using vortex blobs (or vortex particle elements), such that [35]:

$$\overline{\omega}(\mathbf{x},\,t) = \sum_{k=1}^{NV} \frac{\Gamma_k}{\pi\,\sigma_{0_k}^2} \exp\left(-\frac{|\mathbf{x}|^2}{\sigma_{0_k}^2}\right) \tag{8}$$

In Equation (8), $\overline{\omega}$ ($\overline{\omega} = \nabla \times \overline{\mathbf{u}}$) represents the vorticity filtered field; NV is the total number of Lamb discrete vortices necessary to simulate the viscous wake; and Γ_k defines the strength of the k-th vortex blob placed at a pivotal point of a flat panel. The discrete vortices generation is necessary to ensure the no-slip boundary condition Equation (6) in association with the global circulation conservation of the problem. In addition, σ_0 defines the Lamb vortex core, which is measured such as [36,37] (see also Figure 3):

$$\sigma_0 = 1.41421\,\sqrt{\frac{\Delta t}{\text{Re}}} \tag{9}$$

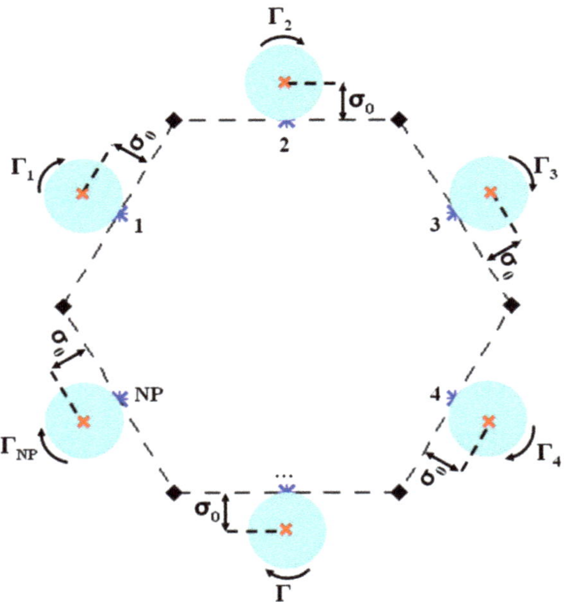

Figure 3. The boundary layer model for a smooth surface.

In Equation (9), the time stepping, Δt, is computed from an estimate of the velocity and advective length of the flow.

As previously described, the potential problem is solved using the panel method through constant-density source distributions [34]. On the other hand, the vorticity effect is included into the problem by taking the curl of Equation (2), which, in association with Equation (1), results in the well-known vorticity transport equation [38]. In two dimensions, the mentioned equation is scalar and the pressure term is eliminated, such as:

$$\frac{\partial\overline{\omega}}{\partial t} + (\overline{\mathbf{u}} \cdot \nabla)\overline{\omega} = \frac{1}{\text{Re}_c}\nabla^2\overline{\omega} \tag{10}$$

being Re_c a local Reynolds number including the local eddy viscosity coefficient, and defined as [25]:

$$Re_{c_i}(t) = \frac{U^* D^*}{\nu + \nu_{t_i}(t)} \tag{11}$$

Equation (10) simulates the vorticity filtered field dynamic, being solved to each vortex blob during each time stepping. Figure 3 sketches the typical generation process of vortex blobs used for a smooth surface. Each vortex blob is positioned tangent to the pivotal point of the panel to ensure the no slip boundary condition of the problem Equation (6), as already commented.

The present roughness model modifies the generation process of vortex particle elements without change the body surface. It is known that the roughness effect on flows is to stimulate the turbulence development. Thus, the mentioned generation process is altered, in order to add an instantaneous inertial effect to each nascent vortex blob. This additional inertial effect works as an injection of momentum into the boundary layer and, as consequence, the flow can support the unfavorable pressure gradient present in the boundary layer for longer, in agreement with the physics involved in this kind of complex flows. Since roughness and turbulence are intimately related, the momentum injection imposed by the present roughness model is obtained from turbulence modeling developed by Lesieur and Métais [33], which was originally implemented in a two-dimensional Lagrangian manner by Alcântara Pereira et al. [25], and with improvements made by Bimbato et al. [37].

Thus, the generation process of vortex blobs is modified considering the turbulent activity around "panel's shedding point" (Figure 2), what is not taken into account for smooth surfaces (Figure 3). The turbulent activity in the vicinity of the shedding point of the i-th panel is computed adopting a semicircle, with radius $\|\mathbf{b}\| = 2\varepsilon - \sigma_0$, and centered on i-th shedding point ($\varepsilon = \varepsilon^*/D^*$ is the mean relative surface roughness). According to Bimbato et al. [14], a set of NR points is placed on the semicircle and the average speed differences, required to compute the second-order velocity structure function of the filtered field Equation (4), are calculated between the center of the semicircle (the shedding point) and the points on it (Figure 4). The second-order velocity function is obtained through the following adapted expression [14]:

$$\overline{F}_{2_i}(t) = \frac{1}{NR} \sum_{w=1}^{NR} \left\| \overline{\mathbf{u}}_{t_i}(\mathbf{x}_i, t) - \overline{\mathbf{u}}_{t_w}(\mathbf{x}_i + \mathbf{b}, t) \right\|_w^2 (1 + \varepsilon) \tag{12}$$

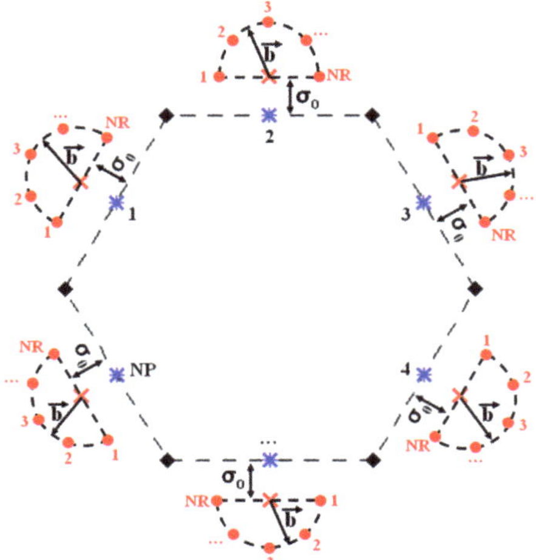

Figure 4. Turbulent activity computation in the vicinity of each shedding point.

In Equation (12), $\bar{u}_{t_i}\,(x_i,\,t)$ is the total velocity on shedding point of the i−th panel (which is located at x_i) at time t, $\bar{u}_{t_w}\,(x_i\,+\,\mathbf{b},\,t)$ is the total velocity at each one of the NR points, defined around the i−th shedding point (that is, on the semicircle of radius $\|\mathbf{b}\| = 2\varepsilon - \sigma_0$) at time t, and $(1 + \varepsilon)$ is idealized to simulate the kinetic energy gain because of the roughness interference. Figure 4 gives us an idea of the turbulent activity computation around the shedding points, which is the velocity fluctuation in the boundary layer.

It is noted from Equation (12) that, at each time stepping, there is a different velocity fluctuation around the shedding point of the i−th panel. Therefore, there is also different local eddy viscosity coefficients associated with these velocity fluctuations, Equations (3) and (4). Thus, the eddy viscosity coefficient at i−th shedding point, at time t, is computed such as:

$$\nu_{t_i}(t) = 0.105\ C_k^{-3/2}\ \sigma_{0_k}\ \sqrt{F_{2_i}(t)} \tag{13}$$

In Equation (13), σ_{0_k} is the vortex core of the k−th vortex blob placed at i−th shedding point Equation (9).

Since the eddy viscosity is associated with each time and with each panel shedding point, a local Reynolds number must be computed at each time stepping, as defined in Equation (11). Therefore, $Re_{c_i}(t)$ represents the Reynolds number modified locally (at i−th shedding point) at time t, and this modification only is computed if the surface roughness effect cannot be neglected, that is, for $\nu_{t_i}(t) \neq 0$.

Equation (9) shows how the Reynolds number of the flow and the Lamb vortex core are related. Thus, if the surface roughness effect is present in the boundary layer flow (if $\nu_{t_i}(t) \neq 0$), the Reynolds number is modified locally Equation (11). The numerical effect is to change the core radius, σ_0, of each nascent vortex blob by satisfying Equation (14) instead of Equation (9). Therefore, the generation process of discrete vortex elements, illustrated in Figure 3, is now modified to include an additional inertial effect to each nascent vortex blob (it can also be interpreted by an increase of strength $\Delta\Gamma$ in Figure 5).

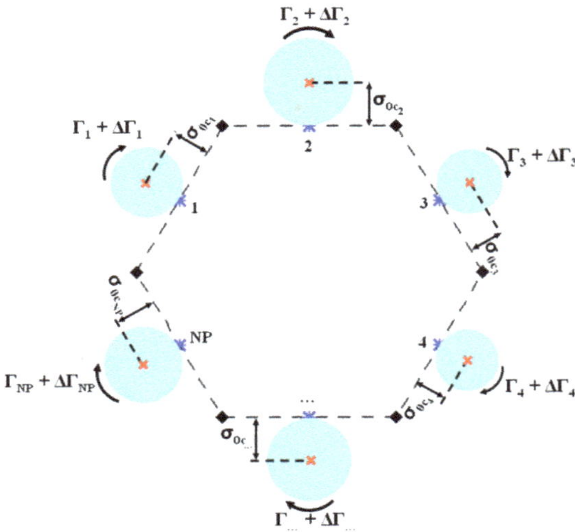

Figure 5. The boundary layer model to simulate a rough surface.

The new Lamb vortex core assumes the following form [14]:

$$\sigma_{0c_k}(t) = 1.41421\ \sqrt{\frac{\Delta t}{Re}\left(1 + \frac{\nu_{t_i}(t)}{\nu}\right)} \tag{14}$$

In Equation (14), $\sigma_{0c_k}(t)$ is the vortex core of k−th vortex blob, placed at i−th shedding point at time t, and modified by surface roughness effect (compare Figure 3 with Figure 5). The roughness model is based only on physics of turbulent flows.

The procedure above described to represent the body surface (sources generation) and the vorticity field (creation of nascent vortex blobs) is mathematically defined by two different kinds of linear algebraic equation systems; their solution is iterative and they are solved using the method of least squares, and can be easily linked with the roughness model, when activated.

The set of discrete vortex elements, positioned at shedding points, is necessary to simulate the vorticity filtered field dynamics according to vorticity Equation (10). With this purpose, Chorin [19] idealized an algorithm that splits the advective-diffusive operator of Equation (10) such as, respectively:

$$\frac{D\overline{\omega}}{Dt} = \frac{\partial\overline{\omega}}{\partial t} + (\overline{\mathbf{u}} \cdot \nabla)\overline{\omega} = 0 \tag{15}$$

$$\frac{\partial\overline{\omega}}{\partial t} = \frac{1}{Re_c}\nabla^2\overline{\omega} \tag{16}$$

Equation (15) implies that the vortex cloud is advected as a set of material particles of the flow in a typical Lagrangian manner.

The solution of the advection problem Equation (15) is given by integrating each vortex blob path equation, and using an explicit Euler scheme, in the following form:

$$\frac{dx_k}{dt} = \overline{\mathbf{u}}_{t_k}(x_k, t), \ k = 1,\ NV \tag{17}$$

In Equation (17), $\overline{\mathbf{u}}_{t_k}(x_k,\ t)$ represents the velocity filtered field computed at position occupied by k−th vortex blob. Thus, the velocity field of the problem is composed by three contributions, i.e., (i) the incident flow (or the uniform flow) velocity, $\overline{ui}(x,t)$; the cylinder surface (using the source panels distribution), $\overline{ub}(x,t)$; the viscous wake (using the vortex cloud), $\overline{uv}(x,t)$.

For the incident flow, shown in Figure 1, its components take the form:

$$\overline{ui_1} = 1 \tag{18}$$

$$\overline{ui_2} = 0 \tag{19}$$

The velocity induced by NP flat panels with constant-density source distribution at position occupied by k−th vortex blob is computed in the following form [34]:

$$\overline{ub_k^n}(x_k, t) = \sum_{i=1}^{NP} \sigma_i c_{ki}^n[x_k(t) - x_i], \ n = 1,\ 2;\ k = 1,\ NV \tag{20}$$

In Equation (20), σ_i is the source strength per panel length and $c_{ki}^n[x_k(t) - x_i]$ represents the n−th component of the velocity induced at the k−th vortex blob by the i-th source flat panel.

The third contribution for the velocity filtered field computation is because of the vortex cloud (i.e., the vortex-vortex interaction, obtained through the Biot-Savart law [35]), which takes the following form:

$$\overline{uv_k^n}(x_k, t) = \sum_{j=1}^{NV} \Gamma_j c_{kj}^n\left[x_k(t) - x_j(t)\right], \ n = 1,\ 2;\ k = 1,\ NV \tag{21}$$

In Equation (20), Γ_j is the strength of the j−th vortex blob and $c_{kj}^n\left[x_k(t) - x_j(t)\right]$ represents the n−th component of the velocity induced at the k−th vortex blob by other vortex blob placed at x_j.

Therefore, the velocity filtered field, $\overline{\mathbf{u}}_t(x,\ t)$, is obtained by the summation of Equations (18)–(20) and its computation is used:

(a) to obtain the position of vortex blobs at the next time stepping, x_k $(t + \Delta t)$, by using the material derivative concept $(D\overline{\omega}/Dt)$;

(b) by roughness model to compute the average speed differences between shedding points and the points on the semicircles (Equation (12) and Figure 4);

(c) by turbulence LES modeling (as will be described below).

Once advective equation Equation (15) has been solved, to properly simulate the vorticity field dynamics, it is necessary to solve, within the same time stepping, the diffusion equation (Equation (16)) to include molecular viscous effect. In the present numerical method, Equation (16) is solved through the random walk method [21], which was originally developed by Chorin [19]. The integral solution of the random walk method is [19]:

$$\overline{\omega}(\mathbf{x},t) = \int_{-\infty}^{\infty} [G(x,\,y,\,t) - G(x-y,\,t)]f(y)dy \tag{22}$$

being:

$$G(x,\,y,\,t) = \frac{1}{\sqrt{\frac{4\,\pi\,t}{Re_c}}} \exp\left(\frac{-(x-y)^2}{\frac{4t}{Re_c}}\right) \tag{23}$$

Therefore, the diffusive displacement of the $k-$th vortex blob is computed, such as:

$$\varsigma_k(t) = \sqrt{\frac{4\,\Delta t}{Re_c}\ln\left(\frac{1}{P}\right)}\left[\cos(2\,\pi\,Q)_x + \sin(2\,\pi\,Q)_y\right] \tag{24}$$

In Equation (24), P and Q are random numbers, which may assume values in the range: $0 < P < 1$ and $0 < Q < 1$.

The presence of the local Reynolds number (Re_c) in Equation (23) indicates that turbulence manifestations must be computed into the diffusion process [25]. The essence of turbulence modeling used in this work consists on calculating the turbulent activity in points of the flow through the average speed differences existing around such points (Equation (4)). Instead of compute these differences between a center of a sphere and points on its surface, as established by Lesieur and Métais [33] for three-dimensional flows, Alcântara Pereira et al. [25] proposed to compute the average speed differences between a center of an annulus and points located between its internal and external radius. Both the center of the annulus and the points between its inner and outer radius are defined for each vortex blob presents in the computational domain simulating the vorticity field dynamics.

The adaptations proposed by Alcântara Pereira et al. [25] let to compute the second-order velocity structure function of the filtered field in the following form:

$$\overline{F}_{2_k} = \frac{1}{N}\sum_{j=1}^{N} \left\|\overline{\mathbf{u}}_{t_k}(x_k) - \overline{\mathbf{u}}_{t_j}\left(x_k + r_j\right)\right\|_j^2 \left(\frac{\sigma_{0_k}}{r_j}\right)^{2/3} \tag{25}$$

In Equation (25), $\overline{\mathbf{u}}_t$ is the total velocity induced on points of interest ($\left|\overline{\mathbf{u}}_t\right| = \overline{ui} + \overline{ub} + \overline{uv}$), N is the number of discrete vortices inside the annulus and r_j defines the distance between the center of annulus (vortex blob under analysis, i.e., the $k-$th particle) and the points located between the internal and the external radius of annulus (the $j-$th particle placed inside it).

A statistical analysis performed by Bimbato et al. [37] concluded that proper internal and external radius of annulus is, respectively: $r_{int} = 0.1\sigma_{0_k}$ and $r_{ext} = 4.0\sigma_{0_k}$ (for smooth surfaces) or $r_{int} = 0.1\sigma_{0c_k}$ and $r_{ext} = 4.0\sigma_{0c_k}$ (for rough surfaces).

The use of Lagrangian vortex method is justified by the ease of work with high Reynolds number simulations (there is no numerical instabilities, which are associated with Eulerian techniques). Furthermore, there is no mesh in the Lagrangian technique and it is kind simple to implement the turbulence LES modeling using velocity differences instead of derivatives Equation (24), which must be closely related to the roughness model Equation (12).

It remains to calculate the aerodynamic coefficients (drag and lift forces). With this purpose, it is used the stagnation pressure definition, such as:

$$\overline{Y}^{*} = \frac{\overline{p}^{*}}{\rho} + \frac{\overline{u}^{*2}}{2} ; \ \overline{u}^{*} = \left| \overline{u}^{*} \right| \tag{26}$$

In Equation (26), $\overline{p}^{*}$ defines the static pressure and $\overline{u}^{*}$ represents the velocity at any point into the fluid domain Ω (Figure 1). Then, the methodology to compute aerodynamic loads starts from a Poisson equation for the pressure, which is solved through the following integral formulation presented by Shintani and Akamatsu [39]:

$$H\overline{Y}_i - \int_{S_1} \overline{Y} \nabla \Xi_i \cdot \mathbf{e}_n dS = \iint_{\Omega} \nabla \Xi_i \cdot (\overline{u} \times \overline{\omega}) d\Omega - \frac{1}{Re} \int_{S_1} (\nabla \Xi_i \times \overline{\omega}) \cdot \mathbf{e}_n dS \tag{27}$$

In Equation (27), the static pressure can be computed at i−th calculation point, being $H = 1.0$ valid for the fluid domain and $H = 1/2$ valid for solid boundary, Ξ represents a fundamental solution of Laplace equation, and $\mathbf{e}_n$ defines the unit vector normal to each flat panel used to discretize the cylinder surface.

Using the static pressure obtained from Equation (27), the drag and lift coefficients are evaluated, such as, respectively:

$$C_D = \sum_{i=1}^{NP} 2\left(\overline{p}_i - p_\infty\right) \Delta S_i \sin \beta_i = \sum_{i=1}^{NP} C_{P_i} \Delta S_i \sin \beta_i \tag{28}$$

$$C_L = -\sum_{i=1}^{NP} 2\left(\overline{p}_i - p_\infty\right) \Delta S_i \cos \beta_i = -\sum_{i=1}^{NP} C_{P_i} \Delta S_i \cos \beta_i \tag{29}$$

In Equations (28) and (29), p_∞ is the reference pressure far from the solid boundary, ΔS_i is the length of the i−th panel, and β_i is the angle of the i−th panel.

Finally, the algorithm implemented essentially consists of eight steps in the following sequence: (i) simultaneous generation of source panels and nascent vortex blobs (the roughness model is used, when activated); (ii) velocity field computation at each vortex blob; (iii) static pressure computation at each pivotal point followed by integrated aerodynamic loads computation; (iv) advection of the vortex cloud; (v) vorticity diffusion (the LES modeling is used); (vi) reflection of vortex blobs, which eventually migrate into the cylinder; (vii) velocity field computation at each pivotal point to restore the boundary conditions of the problem for new generation of source panels and nascent vortex blobs; (viii) advance by time Δt. In the last years, the authors of this paper have made an effort to develop the in-house code by using FORTRAN programming language.

4. Results and Discussion

This Section presents the numerical results obtained for the problem of the two-dimensional, incompressible and unsteady flow past a circular cylinder. The body surface was represented by $NP = 300$ source flat panels. That chosen number for the panels ensures the convergence for the potential solution of the problem and, as consequence, a refined discretization for the vorticity field, which is reflected on the aerodynamic loads computation. In the case of flow around smooth or rough cylinders for the Reynolds number value studied here ($Re = 1.0 \times 10^5$), the drag coefficient is computed only by form (pressure) drag contribution. According to Achenbach [3] measurements, the friction drag contribution does not exceed 2–3% to the total drag. Thus, the friction drag computation has been omitted in this work.

In the present approach, it has been used NR = 21 points on each semicircle defined around each panel shedding point in order to compute the turbulent activity around it (Figure 4); this number is enough to obtain a reasonable value for the average speed differences Equation (12) in consonance with past study reported by Bimbato et al. [14].

In order to obtain good results, the time increment used is $\Delta t = 0.05$ and an explicit Euler time-marching scheme is used Equation (17). All numerical simulations run until dimensionless time of t = 75, which is enough to reach a statistical equilibrium; the mean coefficient values (pressure distribution, drag and lift) are computed between $37.5 \le t \le 75.0$. As previously presented by Oliveira et al. [18], the adopted criterion to compute aerodynamic loads has sufficiently been refined attaining the saturation state of the numerical simulations; this behavior is typical of a Lagrangian technique.

Table 1 presents a summary of some typical experimental results published in the literature and the numerical ones obtained using the present methodology. There is a remarkable scatter among the experimental results, even those referring to the smooth circular cylinder. The scatter among experimental results is because of influencing parameters, especially by the low aspect ratio and high blockage. The cylinder tested by Achenbach [3] presented an aspect ratio of $L^*/D^* = 3.3$ (L^* is the cylinder length) and the wind tunnel blockage was $D^*/B^* = 0.17$ (B^* is the distance between tunnel walls). The cylinder tested by Achenbach and Heinecke [4] presented an aspect ratio of $L^*/D^* = 6.75$ and the wind tunnel blockage was $D^*/B^* = 0.17$, while Zhou et al. [12] used a configuration of cylinder with aspect ratio of $L^*/D^* = 10$. Thus, the comparison among experimental results and the present numerical results is not so fair, as previously discussed in Section 1. Unfortunately, there is a lack of numerical data of aerodynamic forces for the flow past a rough cylinder at upper subcritical Reynolds number of $Re = 1.0 \times 10^5$.

Table 1. Summary for the drag coefficient at $Re = 1.0 \times 10^5$.

$\varepsilon = \varepsilon^*/D^*$	Achenbach [3]	Achenbach and Heinecke [4]	Zhou et al. [12]	Present Simulation
Smooth	1.30	1.48	1.14	1.223
0.075%	-	1.44	-	-
0.10%	-	-	-	1.188
0.11%	1.10	-	-	-
0.20%	-	-	-	1.133
0.30%	-	1.25	-	-
0.45%	0.82	-	-	1.025
0.70%	-	-	-	1.071
0.90%	1.10	1.18	-	-
2.00%	-	-	0.92	-

Despite this, it is undeniable that the present numerical results have the same tendency as the experimental results of Achenbach [3] (Table 1). The roughness model used here is able to predict supercritical flow features from a subcritical Reynolds number flow simulation. This conclusion comes from the comparison between Figure 6a,b.

The present numerical results show that, for small superficial roughness ($0.00 < \varepsilon \le 0.10\,\%$), the drag coefficient suffers a very small reduction (about 2.9%), which occurs in the subcritical regime (Figure 6a). With the progressive increase in surface roughness ($0.20\% \le \varepsilon \le 0.45\%$), the roughness model injects momentum on the boundary layer in a manner that flows with higher Reynolds numbers are simulated; it is remarkable the drag coefficient sudden drops until reach a minimum value (Figure 6b), that is a typical feature of critical flows. The mean drag reduction is about 16.2%. For even greater roughness ($\varepsilon > 0.45\,\%$), the drag coefficient grows up gradually (about 4.5%), which is the main feature of the supercritical flow regime (Figure 6a).

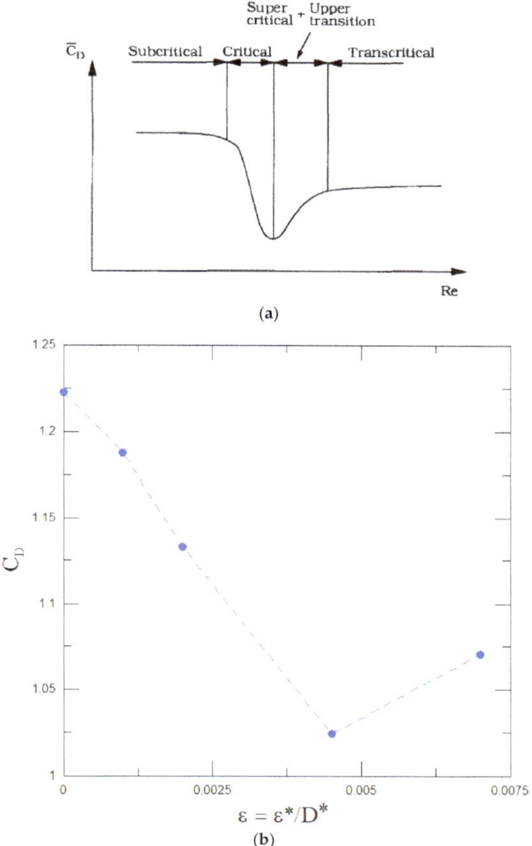

Figure 6. Flow regimes described by drag coefficient: (**a**) Classical representation of $\overline{C}_D = \overline{C}_D$ (Re) curve for a cylinder (reproduced from [40] with permission; copyright © 2006 by World Scientific Publishing Co. Pte. Ltd.); (**b**) Form of $\overline{C}_D = \overline{C}_D$ (ε) curve obtained by roughness model for a fixed Reynolds number (Re $= 1.0 \times 10^5$).

Although the percentage drop in the drag coefficient measured by Achenbach [3] is slightly more than double that calculated in the present work, most of the features of transitional flows are qualitatively captured by present methodology. Table 2 shows that boundary layer separation point agree well with the drag coefficient behavior pointed out in Table 1 and Figure 6b, what means the separation point moves downstream with drag reduction (critical flow regime) and upstream with drag gradual increase (supercritical flow regime).

Table 2. Upper side angular position of boundary layer separation, θ_{sep} (Re = 1.0×10^5).

$\varepsilon = \varepsilon^*/D^*$	Achenbach [3]	Present Simulation
Smooth	77°	81.0°
0.10%	-	84.0°
0.20%	-	84.0°
0.45%	102°	93.0°
0.70%	-	89.4°
0.90%	99°	-

Furthermore, Figure 7 indicates that the lift coefficient,C_L, oscillates regularly because of the periodic vortex shedding mechanism [41], and the lift force variation of the smooth cylinder is greater than that of the rough cylinder, which agrees with experimental tests conducted by Zhou et al. [12]; the dash-dot lines in Figure 7 indicates the mean amplitude of lift force. In the present work, no analysis regarding the vortex shedding frequency is presented, however, a detailed discussion can be easily recuperated [14,17,18].

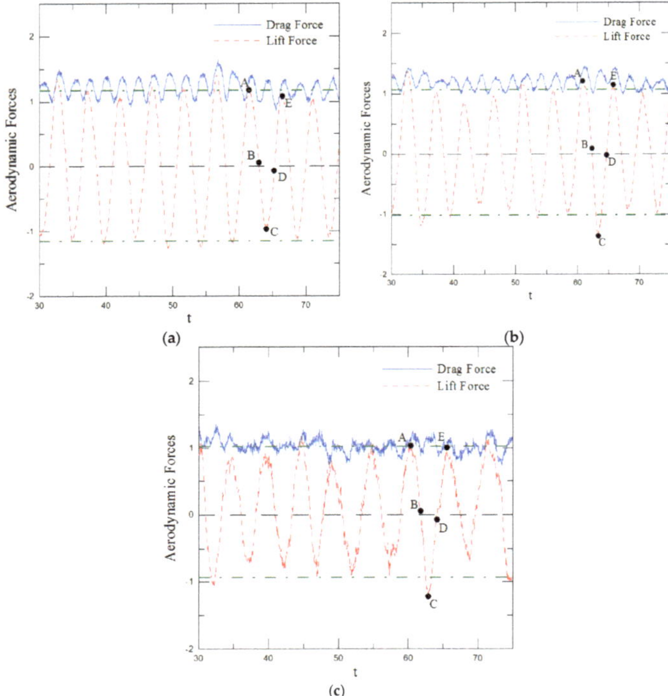

Figure 7. Temporal history for drag and lift coefficients (Re $= 1.0 \times 10^5$): (**a**) Smooth; (**b**) $\varepsilon = 0.10\%$; (**c**) $\varepsilon = 0.45\%$.

Figure 8 presents the pressure distributions around the cylinder surface at time instants represented by points A, B, C, D and E; these points are also shown in Figure 7. It can be noticed that the surface roughness effect causes irregular disturbances on instantaneous pressure distribution curves, which is reflected on time history of drag and lift forces.

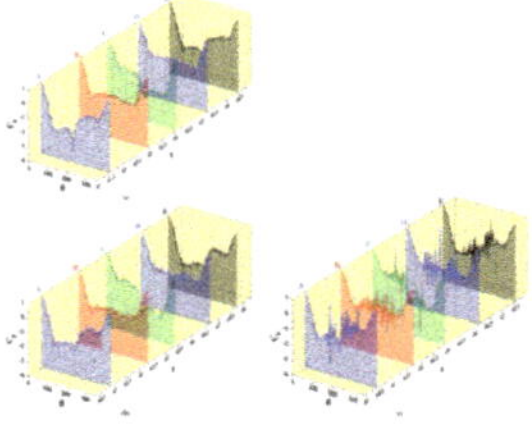

Figure 8. Pressure distributions for particular time instants of the flow past a circular cylinder (Re $= 1.0 \times 10^5$): (**a**) Smooth; (**b**) $\varepsilon = 0.10\%$; (**c**) $\varepsilon = 0.45\%$.

The mentioned disturbances occur because of inertial effect manifesting on the flow, which is provoked by roughness model. Thus, the boundary layer presents more momentum supporting the unfavorable pressure gradient, as expected, and, as consequence, the flow separation is delayed. This was shown in the joint analysis of Figure 6b and Table 2; now it can also be verified with the analysis of Figure 9.

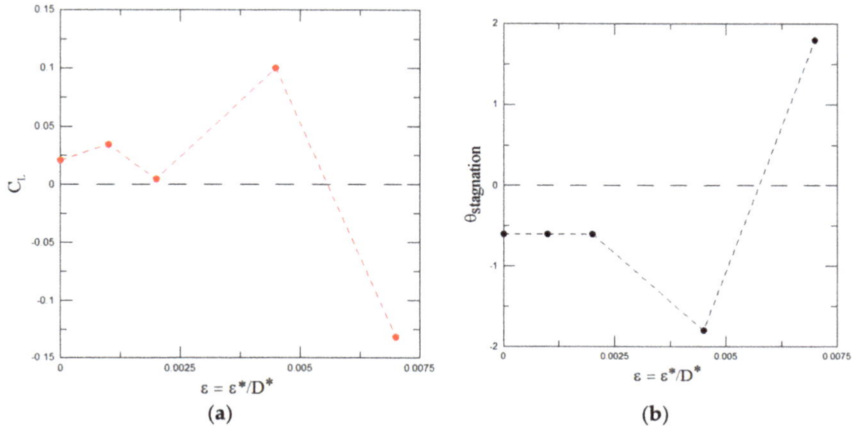

Figure 9. Separation bubble identification (Re $= 1.0 \times 10^5$): (**a**) Mean lift coefficient; (**b**) Stagnation point.

In Figure 9a, the non-zero lift force for the smooth cylinder ($\overline{C}_L = 0.021$) and for the less roughened cylinders ($\overline{C}_L = 0.035$ for $\varepsilon = 0.10\%$ and $\overline{C}_L = 0.005$ for $\varepsilon = 0.20\%$) can be attributed to numerical rounding errors (panels method and vortex cloud contributions). It can be noted that the stagnation point of these three tests is located at $\theta_{stag} = 0.6°$; this happens because the discretization of the cylinder using the panel method starts with $\theta = 0°$ (Figure 1) and the panel pivotal point is placed at $\theta = 0.6°$. It is important to emphasize that all the computations are done on pivotal points (Figure 2). So, this is the reason why $\theta_{stag} \neq 0$ for the three cases mentioned.

On the other hand, it is notorious not only the non-zero lift force for the rougher cylinders but its direction change ($\overline{C}_L = +0.100$ and $\theta_{stag} = -1.8°$ for $\varepsilon = 0.45\%$; $\overline{C}_L = -0.132$ and $\theta_{stag} = +1.8°$ for $\varepsilon = 0.70\%$), see Figure 9a,b. The boundary layer transition is the only explanation of this occurrence on a symmetrical bluff body, like the circular cylinder. The involved physics indicates that first an asymmetric flow is caused because the boundary layer becomes turbulent at one side of the cylinder; furthermore, a separation bubble and a non-zero lift force can be identified. On the other side of the cylinder is generated another separation bubble, when the boundary layer becomes turbulent on that side. According to Kamiya et al. [6], a fully symmetric flow is not reach as soon as the second separation bubble is formed. They measured $\overline{C}_L = 0.2$ and $\theta_{stag} = +2°$ when the first separation bubble was formed, followed by $\overline{C}_L = 0.0$ and $\theta_{stag} = 0°$ at higher Reynolds number (Re $\geq 7.0 \times 10^5$). The critical flow regime is so unstable, and the lift force changing direction was also captured by Kamiya et al. [6] by using a smooth cylinder.

One more feature of the complex transitional flow is captured by the present methodology. According to Bearman [5], the drag coefficient is approximately equal to the measured base pressure coefficient for three representative Reynolds numbers, i.e., Re $= 2.0 \times 10^5$ (no separation bubble), Re $= 3.7 \times 10^5$ (one separation bubble) and Re $= 4.0 \times 10^5$ (two separation bubbles). Figure 10 shows the comparison between drag and base pressure coefficients, calculated by the present Lagrangian approach.

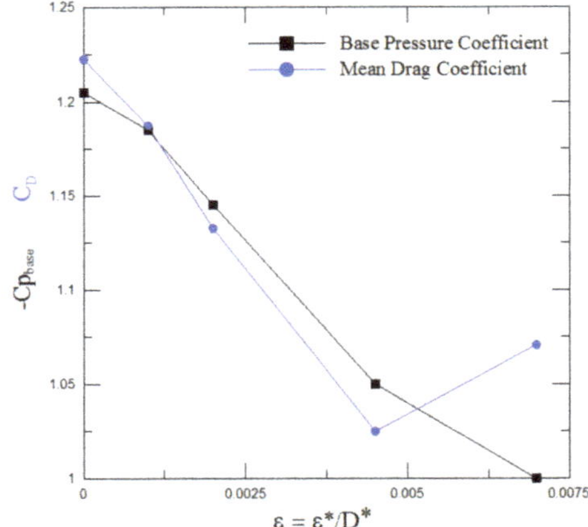

Figure 10. Comparison between computed results of base pressure and mean drag coefficients
(Re $= 1.0 \times 10^5$).

It is remarkable that the present methodology agrees with the observations given by Bearman [5]
(using the $L^*/D^* = 12$ and $D^*/B^* = 0.06$ configuration) for the critical flow regime. As expected, there is a
disagreement between the computed values of base pressure and drag coefficients for the supercritical
flow regime, because of increasing in drag coefficient.

In addition, Figure 11 presents the time-averaged pressure distribution for the flow around a
cylinder with different surface roughness influences, where the experimental result was reported by
Blevins [42] for smooth cylinder. It can be observed only small differences in the base pressure between
the less rough cylinder and the smooth cylinder. As can be identified, the numerical result for the
rough cylinder ($\varepsilon = 0.45\%$) clearly shows a higher increase in the base pressure, which physically
agrees with the higher drag reduction (Figure 6b).

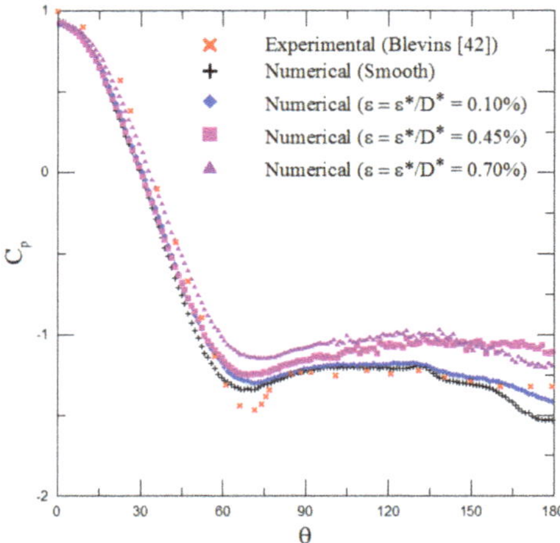

Figure 11. Time-averaged pressure distribution around the circular cylinder surface (Re $= 1.0 \times 10^5$).

Finally, Figure 12 illustrates the viscous wake developed downstream the cylinder in the end of
numerical simulations (at $t = 75$). The blue points in the same figure represent instantaneous vortex
blob distributions; they indicate that the vortical structures for $\varepsilon = 0.45\%$ are narrower than the

other ones, which is in accordance to what is expected from critical flow regimes. That conclusion is supported by upper and lower separation points behavior for $\varepsilon = 0.45\%$, which are 93° (Table 2) and 85.8°, respectively.

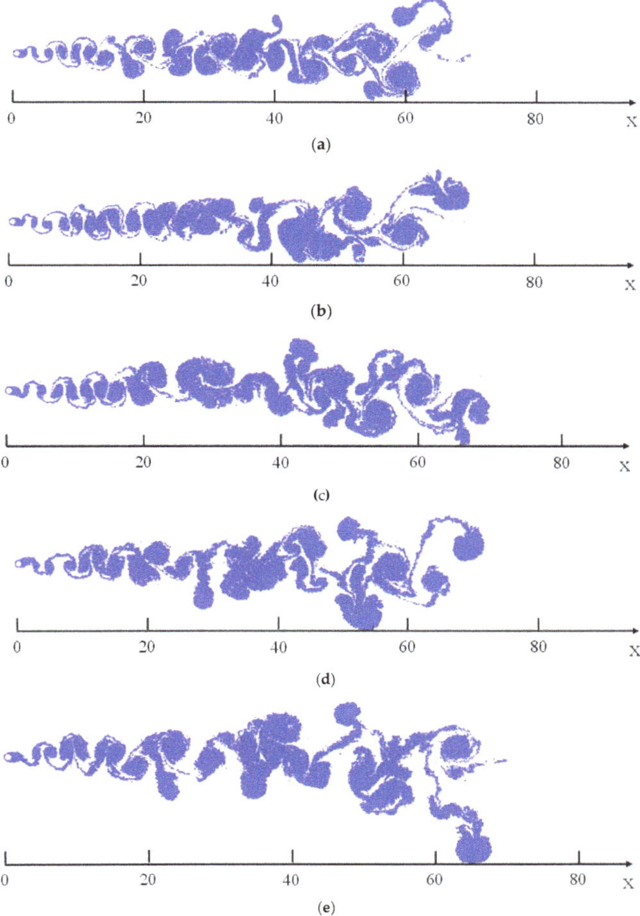

Figure 12. Final position of the viscous wake at t = 75 (Re $= 1.0 \times 10^5$): (**a**) Smooth; (**b**) $\varepsilon = 0.10\%$; (**c**) $\varepsilon = 0.20\%$; (**d**) $\varepsilon = 0.45\%$; (**e**) $\varepsilon = 0.70\%$.

5. Conclusions

This paper describes a two-dimensional Lagrangian vortex method blended with both LES and roughness models to capture changes in a bluff body aerodynamics. It can be observed that the roughness model employed in the present work is classified as a passive method of vortex shedding control. The passive control method implies to modify the body surface, which is reflected on its aerodynamic loads behaviour. In fact, the roughness model presented here acts as a passive method modifying the boundary layer flow by an injection of momentum into it (which occurs on rough surfaces), but without changing the discretization of the body surface.

In the present approach, the vorticity field is discretized and represented by a cluster of vortex blobs, which are generated during each time stepping from the body surface. The potential theory is adapted to compute the circular cylinder surface, which is discretized using flat panels with constant-density source distribution over them. The aerodynamic loads are calculated through an integral formulation derived from a Poisson equation for the pressure, where the instantaneous vorticity field contributes to the computation. The chosen number of flat panels ensures the convergence of the solution for the potential velocity field and, as consequence, establishes the number of nascent vortex blobs. As the non-dimensional time runs, the number of vortex blobs increases and thus satisfies the vorticity transport equation forming the von-Kármán vortex street. In an upper subcritical Reynolds flow, the drag coefficient amplitudes are significantly lesser then the lift ones (Figure 7a). This behaviour indicates that the numerical simulations have captured the vortex shedding mechanism properly; it is important to observe that the saturation state of the present numerical simulations was attained and it is supported by previous discussions given by Oliveira et al. [18], since the numerical approach is the same.

In regarding of the rough circular cylinder, the numerical results show that the present methodology is able to capture, even using two-dimensional simulations, many attributes of this complex flow such as: (i) the drag crisis followed by a gradual increase on the drag force; (ii) the non-zero lift force, since according to Zdravkovich [10] the low and high pressure on two sides of the cylinder interchange in different runs, because the side at which the separation is turbulent switches from one side to another occasionally; and (iii) the occurrence of asymmetric separation bubbles generation. The asymmetries are part of the physics involved in critical Reynolds number flows around a circular cylinder; that behavior was previously described by Zdravkovich [10], whom dedicated all his research to investigate different aspects of flows around cylinders.

In order to obtain the results showed here, the present work ran simulations with roughness heights of $\varepsilon = 0.00\%$ (smooth case), $\varepsilon = 0.10\%$, $\varepsilon = 0.20\%$, $\varepsilon = 0.45\%$ and $\varepsilon = 0.70\%$. The roughness heights of $\varepsilon = 0.10\%$ and $\varepsilon = 0.20\%$ were performed to show that their effect does not change the flow patterns. Further, the tests jumped for $\varepsilon = 0.45\%$ (+0.25%) and then for $\varepsilon = 0.70\%$ (+0.25%). The critical ($\varepsilon = 0.45\%$) and upper transition ($\varepsilon = 0.70\%$) Reynolds number flows are so unstable, that it could be difficult to physically explain the results on a higher resolution area around $\varepsilon = 0.45\%$.

Although the lift force behavior is an important feature of the critical/supercritical flow regimes, studies that focus on this force are scarce, which valorizes the present numerical results. The results for both predictions of boundary layer separation and stagnation point displacement also help us to report the sensitivity of the roughness model. Furthermore, each experimental study available in the literature uses a different technique to represent the surface roughness effect (wires, spheres, small holes, among others), and each technique is subject to different influencing parameters (free stream turbulence, aspect ratio, blockage and surface texture), which produces a great scatter in the measurements. As consequence, in most of situations it is not appropriate to compare two different experimental results or even an experimental and a numerical one.

Finally, the authors also intend to adapt the present vortex code to study other practical engineering situations, such as problems involving wake interference [43], ground effect mechanisms [17,18,27,28], vortex-induced vibrations [13,26] and thermal effect [44,45]. The present paper also aims to lay the theoretical foundation of the present numerical method for further algorithmic extensions to simulations of such fluid flows in three dimensions.

Author Contributions: Conceptualization, A.M.B., L.A.A.P. and M.H.H.; data curation, A.M.B.; formal analysis, A.M.B., L.A.A.P. and M.H.H.; funding acquisition, L.A.A.P.; investigation, A.M.B., L.A.A.P. and M.H.H.; methodology, A.M.B., L.A.A.P. and M.H.H.; project administration, A.M.B. and L.A.A.P.; resources, A.M.B. and L.A.A.P.; software using FORTRAN, A.M.B. and L.A.A.P.; validation, A.M.B.; visualization, A.M.B.; supervision, L.A.A.P.; writing—Original draft, A.M.B.; and writing—review and editing, A.M.B. and L.A.A.P. All authors have read and agreed to the published version of the manuscript.

Funding: This research was funded by FAPEMIG, grant number APQ-02175-14.

Conflicts of Interest: The authors declare no conflict of interest.

References

1. Bearman, P.W. Vortex shedding from oscillating bluff bodies. *Annu. Rev. Fluid Mech.* **1984**, *16*, 195–222. [CrossRef]
2. Tani, I. Low speed flows involving bubble separations. *Prog. Aerosp. Sci.* **1964**, *5*, 70–90. [CrossRef]
3. Achenbach, E. Influence of surface roughness on the cross-flow around a circular cylinder. *J. Fluid Mech.* **1971**, *46*, 321–335. [CrossRef]
4. Achenbach, E.; Heinecke, E. On vortex shedding from smooth and rough cylinders in the range of Reynolds numbers 6×10^3 to 5×10^6. *J. Fluid Mech.* **1981**, *109*, 239–251. [CrossRef]
5. Bearman, P.W. On vortex shedding from a circular cylinder in the critical Reynolds number regime. *J. Fluid Mech.* **1969**, *37*, 577–585. [CrossRef]
6. Kamiya, N.; Suzuki, S.; Nishi, T. On the Aerodynamic Force acting on a Circular Cylinder in the Critical Range of the Reynolds Number. In Proceedings of the AIAA Meeting, New Orleans, LA, USA, 17 January 1979; pp. 79–1975.
7. Guven, O.; Farell, C.; Patel, V.C. Surface roughness effects on the mean flow past circular cylinders. *J. Fluid Mech.* **1980**, *98*, 673–701. [CrossRef]
8. Fage, A.; Warsap, J.H. The Effects of Turbulence and Surface Roughness on the Drag of a Circular Cylinder. *Brit. Aeronaut. Res. Counc. Rep. Memo* **1929**, *1*, 1283.
9. Szechenyi, E. Supercritical Reynolds number simulation for two dimensional flow over circular cylinders. *J. Fluid Mech.* **1975**, *70*, 529–542. [CrossRef]
10. Zdravkovich, M.M. *Flow around Circular Cylinders: Fundamentals*; Oxford University Press: Oxford, UK, 1997; Volume 1.
11. Van Hinsberg, N.P. The Reynolds number dependency of the steady and unsteady loading on a slightly rough circular cylinder: From subcritical up to high transcritical flow state. *J. Fluids Struct.* **2015**, *55*, 526–539. [CrossRef]
12. Zhou, B.; Wang, X.; Gho, W.M.; Tan, S.K. Force and flow characteristics of a circular cylinder with uniform surface roughness at subcritical Reynolds numbers. *Appl. Ocean Res.* **2015**, *49*, 20–26. [CrossRef]
13. Gao, Y.; Zong, Z.; Zou, L.; Takagi, S.; Jiang, Z. Numerical simulation of vortex-induced vibration of a circular cylinder with different surface roughnesses. *Mar. Struct.* **2018**, *57*, 165–179. [CrossRef]
14. Bimbato, A.M.; Alcântara Pereira, L.A.; Hirata, M.H. Development of a new Lagrangian vortex method for evaluating effects of surfaces roughness. *Eur. J. Mech. B Fluids* **2019**, *74*, 291–301. [CrossRef]
15. Du, X.; Lin, W.; Wu, G.; Daichin, J.B. Effects of surface roughness on the wake-induced instabilities of two circular cylinders. *J. Fluids Struct.* **2019**, *91*, 102738. [CrossRef]
16. Ma, W.; Liu, Q.; Macdonald, J.H.G.; Yan, X.; Zheng, Y. The effect of surface roughness on aerodynamic forces and vibrations for a circular cylinder in the critical Reynolds number range. *J. Wind Eng. Ind. Aerodyn.* **2019**, *187*, 61–72. [CrossRef]
17. Alcântara Pereira, L.A.; de Oliveira, M.A.; de Moraes, P.G.; Bimbato, A.M. Numerical experiments of the flow around a bluff body with and without roughness model near a moving wall. *J. Braz. Soc. Mech. Sci. Eng.* **2020**, *42*, 129. [CrossRef]
18. Oliveira, M.A.; Moraes, P.G.; Andrade, C.L.; Bimbato, A.M.; Alcântara Pereira, L.A. Control and Suppression of Vortex Shedding from a Slightly Rough Circular Cylinder by a Discrete Vortex Method. *Energies* **2020**, *13*, 4481. [CrossRef]
19. Chorin, A.J. Numerical study of slightly viscous flow. *J. Fluid Mech.* **1973**, *57*, 785–796. [CrossRef]
20. Leonard, A. Vortex methods for flow simulation. *J. Comput. Phys.* **1980**, *37*, 289–335. [CrossRef]
21. Lewis, R.I. Vortex Element Methods, The Most Natural Approach to Flow Simulation—A Review of Methodology with Applications. In Proceedings of the 1st International Conference on Vortex Methods, Kobe, Japan, 4–5 November 1999; pp. 1–15.
22. Kamemoto, K. On contribution of advanced vortex element methods toward virtual reality of unsteady vortical flows in the new generation of CFD. *J. Braz. Soc. Mech. Sci. Eng.* **2004**, *26*, 368–378. [CrossRef]
23. Faure, T.M.; Leogrande, C. High angle-of-attack aerodynamics of a straight wing with finite span using a discrete vortex method. *Phys. Fluids* **2020**, *32*, 104109. [CrossRef]

24. Sessarego, M.; Feng, J.; Ramos-García, N.; Horcas, S.G. Design optimization of a curved wind turbine blade using neural networks and an aero-elastic vortex method under turbulent inflow. *Renew. Energ.* **2020**, *146*, 1524–1535. [CrossRef]

25. Alcântara Pereira, L.A.; Hirata, M.H.; Silveira Neto, A. Vortex method with turbulence sub-grid scale modeling. *J. Braz. Soc. Mech. Sci. Eng.* **2003**, *25*, 140–146.

26. Hirata, M.H.; Alcântara Pereira, L.A.; Recicar, J.N.; Moura, W.H. High Reynolds number oscillations of a circular cylinder. *J. Braz. Soc. Mech. Sci. Eng.* **2008**, *30*, 304–312. [CrossRef]

27. Bimbato, A.M.; Alcântara Pereira, L.A.; Hirata, M.H. Simulation of viscous flow around a circular cylinder near a moving ground. *J. Braz. Soc. Mech. Sci. Eng.* **2009**, *31*, 243–252. [CrossRef]

28. Bimbato, A.M.; Alcântara Pereira, L.A.; Hirata, M.H. Study of the vortex shedding flow around a body near a moving ground. *J. Wind Eng. Ind. Aerodyn.* **2011**, *99*, 7–17. [CrossRef]

29. Ricciardi, T.R.; Wolf, W.R.; Bimbato, A.M. Fast multipole method applied to Lagrangian simulations of vortical flows. *Commun. Nonlinear Sci. Numer. Simul.* **2017**, *51*, 180–197. [CrossRef]

30. Giachetti, A.; Bartoli, G.; Mannini, C. Two-dimensional study of a rectangular cylinder with a forebody airtight screen at a small distance. *J. Wind Eng. Ind. Aerodyn.* **2019**, *189*, 11–21. [CrossRef]

31. Behara, S.; Chandra, V.; Ravikanth, B. Flow-induced oscillations of three tandem circular cylinders in a two-dimensional flow. *J. Fluids Struct.* **2019**, *91*, 1–16. [CrossRef]

32. Smagorinsky, J. General circulation experiments with the primitive equations. *Mon. Weather Rev.* **1963**, *91*, 99–164. [CrossRef]

33. Lesieur, M.; Métais, O. New trends in large eddy simulation of turbulence. *Rev. Fluid Mech.* **1996**, *28*, 45–82. [CrossRef]

34. Katz, J.; Plotkin, A. *Low Speed Aerodynamics: From Wing Theory to Panel Methods*; McGraw Hill Inc.: New York, NY, USA, 1991.

35. Kundu, P.K. *Fluid Mechanics*; Academic Press: Cambridge, MA, USA, 1990.

36. Mustto, A.A.; Hirata, M.H.; Bodstein, G.C.R. Discrete Vortex Method Simulation of the Flow around a Circular Cylinder with and without Rotation. In Proceedings of the AIAA, Anaheim, CA, USA, 28–30 September 1998; pp. 98–2409.

37. Bimbato, A.M.; Alcântara Pereira, L.A.; Hirata, M.H. Corrected Lagrangian LES Model for Vortex Method. In Proceedings of the 8th Spring School on Transition and Turbulence, São Paulo, Brazil, 24–28 September 2012.

38. Batchelor, G.K. *An Introduction to Fluid Dynamics*; Cambridge University Press: Cambridge, MA, USA, 1967.

39. Shintani, M.; Akamatsu, T. Investigation of two-dimensional discrete vortex method with viscous diffusion model. *CFD J.* **1994**, *3*, 237–254.

40. Sumer, B.M.; Fredsøe, J. Hidrodynamics around Cylindrical Structures. *Adv. Ser. Ocean Eng.* 2006.

41. Gerrard, J.H. The mechanics of the formation region of vortices behind bluff bodies. *J. Fluid Mech.* **1966**, *25*, 401–413. [CrossRef]

42. Blevins, R.D. *Applied Fluid Dynamics Handbook*; Van Nostrand Reinhold Co.: Washington, DC, USA, 1984.

43. Sumner, D.; Reitenbach, H.K. Wake interference effects for two finite cylinders: A brief review and some new measurements. *J. Fluids Struct.* **2019**, *89*, 25–38. [CrossRef]

44. Dierich, F.; Nikrityuk, P.A. A numerical study of the impact of surface roughness on heat and fluid flow past a cylindrical particle. *Int. J. Therm. Sci.* **2013**, *65*, 92–103. [CrossRef]

45. Los Reis, J.H.; Alcântara Pereira, L.A. Particle-particle interactions in parallel computations. In Proceedings of the 16th Brazilian Congress of Thermal Sciences and Engineering, Vitoria, Brazil, 7–10 November 2016.

Publisher's Note: MDPI stays neutral with regard to jurisdictional claims in published maps and institutional affiliations.

 *energies*

Article

Control and Suppression of Vortex Shedding from a Slightly Rough Circular Cylinder by a Discrete Vortex Method

Marcos André de Oliveira [1], **Paulo Guimarães de Moraes** [1], **Crystianne Lilian de Andrade** [1], **Alex Mendonça Bimbato** [2] **and Luiz Antonio Alcântara Pereira** [1,*]

[1] Mechanical Engineering Institute, Federal University of Itajubá (UNIFEI), Itajubá MG 37.500-903, Brazil; marcos.oliveira@uft.edu.br (M.A.d.O.); pauloguimor@gmail.com (P.G.d.M.); crystiannelilian@yahoo.com.br (C.L.d.A.)

[2] School of Engineering, São Paulo State University (UNESP), Guaratinguetá SP 12.516-410, Brazil; alex.bimbato@unesp.br

* Correspondence: luizantp@unifei.edu.br; Tel.: +55-21-35-3629-1041

Received: 31 July 2020; Accepted: 26 August 2020; Published: 31 August 2020

Abstract: A discrete vortex method is implemented with a hybrid control technique of vortex shedding to solve the problem of the two-dimensional flow past a slightly rough circular cylinder in the vicinity of a moving wall. In the present approach, the passive control technique is inspired on the fundamental principle of surface roughness, promoting modifications on the cylinder geometry to affect the vortex shedding formation. A relative roughness size of $\varepsilon^*/d^* = 0.001$ (ε^* is the average roughness and d^* is the outer cylinder diameter) is chosen for the test cases. On the other hand, the active control technique uses a wall plane, which runs at the same speed as the free stream velocity to contribute with external energy affecting the fluid flow. The gap-to-diameter varies in the range from $h^*/d^* = 0.05$ to 0.80 (h^* is the gap between the moving wall and the cylinder bottom). A detailed account of the time history of pressure distributions, simultaneously investigated with the time evolution of forces, Strouhal number behavior, and boundary layer separation are reported at upper-subcritical Reynolds number flows of $Re = 1.0 \times 10^5$. The saturation state of the numerical simulations is demonstrated through the analysis of the Strouhal number behavior obtained from temporal history of the aerodynamic loads. The present work provides an improvement in the prediction of Strouhal number than other studies no using roughness model. The aerodynamic characteristics of the cylinder, as well as the control of intermittence and complete interruption of von Kármán-type vortex shedding have been better clarified.

Keywords: bluff body; roughness model; Venturi effect; suppression hybrid control; Lagrangian description

---

1. Introduction

In the literature, "bluff body" is defined as being a structure that when immerse in a fluid flow will present significant proportion of its surface generating separated flow. This idea is also associated with characteristics of the flow around the body, especially the two shear layers of opposite signals formed from the separation points of the body [1]. Since 1900, numerous analytical, numerical, and experimental investigations have been conducted to study bluff body aerodynamics on either two-dimensional or axisymmetric shapes; the two most common bodies being the circular cylinder and the sphere. The more relevant results have contributed to develop important researches in aerospace/aeronautical, civil, marine, mechanical, and computer engineering.

The present paper aims to contribute with more discussions concerning the surface roughness effect on bluff body aerodynamics. An important practical engineering application of the flow around cylindrical structures is that of the fluid–elastic interaction between the flow and the structure exciting the body into flow-induced vibrations (FIV) [2–6]. Undoubtedly, the surface roughness effect must be used to control this non-linear hydrodynamic phenomenon. This is one of important motivations for future extension of our present research, which will also include the study of slender body aerodynamics [7]. Another one is planned to investigate heat transfer behavior in mixed convection of a fluid by a temperature particles method [8].

In the context of the present work, the literature has reported studies of surface roughness effect on flow past a bluff body with low frequency [2,5,6,9–19]. It is important to mention that numerical analysis of rough bluff body aerodynamics near a ground plane is very limited, which valorizes the recent methodology proposed by Alcântara Pereira et al. [10] and classified such as "hybrid control technique of vortex shedding" [20].

In fact, control and suppression of vortex shedding from a bluff body has been considered as one of the most important research areas in the field of aerodynamics and hydrodynamics applications, such as vibration of pipelines, interactions of currents and wave with offshore structures, suspension bridges and chimneys near tall buildings.

There are different situations where the vortex shedding may cease, and, of particular interest here, is that of a circular cylinder in the vicinity of a ground plane. To the best of our knowledge, the ground effect is governed by three mechanisms [21–28]: (i) The wake interference because of the intertwine of the body wake and the boundary layer formed on the ground, the latter is less influent despite several intensive studies reported so far. (ii) The three-dimensional effect, which presents a momentum transfer in the axial direction of the body leading to a lower drag force value as compared to the two-dimensional results. (iii) The blockage effect (or Venturi effect), which contributes to appearing big and small peaks during the temporal evolution of the drag curve. These peaks have been identified within the large-gap regime, i.e., $h^*/d^* > 0.40$, by Bimbato et al. [22] using a discrete vortex method implemented with Large Eddy Simulation (LES) theory. The large-gap regime is characterized by the presence of strong vortical structures, which are generated at the rear part of a bluff body [25]. In Section 3.2, these peculiar peaks' behavior on the drag curve will be again studied, now using the present numerical method, to contribute on discussion of the roughness model sensitivity. The applicability of the Venturi effect in aerodynamic models has been discussed in the literature [29,30].

Within the scenario above described, Roshko et al. [26] reported the aerodynamic forces behavior for the smooth circular cylinder in the vicinity of a fixed wall using a wind tunnel at high Reynolds number flow of $Re = 2.0 \times 10^4$. As the cylinder came close to the ground plane, it was reported that the drag force rapidly decreased; on the contrary, the lift force increased.

Zdravkovich [27] also reported aerodynamic forces behavior for the smooth circular cylinder near a fixed ground at high Reynolds numbers flows in the range of $4.8 \times 10^4 < Re < 3.0 \times 10^5$. As the gap-to-diameter ratio h^*/d^*, reduced to less than the thickness of the boundary layer $\delta^*/d^* (\delta^*$ is the boundary layer thickness) on the ground plane, it was identified a rapid decrease in the drag force. The drag variation was dominated by h^*/δ^* and the ratio h^*/d^* was less dominant. It was observed that the state of the boundary layer could interfere in the lift force, although the boundary layer thickness was less influent.

In other relevant research, Zdravkovich [28] investigated the drag force behavior of the smooth circular cylinder near a ground plane. The wall plane ran at the same speed as the free stream velocity in an upper-subcritical Reynolds number flow of $Re = 2.5 \times 10^5$ into the critical flow regime. The experimental results showed contrast to all previous studies using fixed ground, e.g., [26,27], because practically no boundary layer was generated from the moving wall. Interestingly, the expected decrease of the drag force, as the ratio h^*/d^* decreased, did not appear. That behavior was attributed to the non-existence of the boundary layer formed on the ground or the high Reynolds number, or any other influencing factors, such as surface texture and structural vibration.

Nishino [25] reported experimental results of the flow around the smooth circular cylinder with an aspect ratio of 8.33 in a wind tunnel. Two upper-subcritical Reynolds numbers of $Re = 0.4 \times 10^4$ and 1.0×10^5 were investigated during his experiments. The cylinder with and without end plates configurations were near and parallel to a wall plane running at the same speed as the free stream velocity. The moving wall effect eliminated the less influent effects of boundary layer formed on the ground and, therefore, the experimental results contributed to clarify the fundamental mechanisms of ground effect in more details. The experimental study also produced new perceptions into the physics of ground effect, and still nowadays serves as database to support numerical investigations. Of importance for the present work, is that, for the cylinder with end plates, the oil flow patterns were observed to be approximately two-dimensional. In contrast to the cylinder near a fixed ground configuration [26], the drag force rapidly decreased, as the ratio h^*/d^* decreased to less than 0.50 and became constant for h^*/d^* of less than 0.35.

Bimbato et al. [21,22] implemented an algorithm of the discrete vortex method with LES modeling to study the two-dimensional flow around the circular cylinder near a moving wall. Their numerical strategy was to represent the ground plane motion using a plane wall with no vorticity generation on it. The numerical results for an upper-subcritical Reynolds number of $Re = 1.0 \times 10^5$ presented a good agreement with the experimental results reported by Nishino [25] using the cylinder with end plates configuration. The authors concluded that the Venturi effect almost completely suppressed the vortex shedding from the cylinder placed closer to the ground plane. Furthermore, the drag force decreased as consequence of the suppression.

In a recent paper, Alcântara Pereira et al. [10] proposed a hybrid control technique of vortex shedding, combining passive and active controls, to study the flow past the rough circular cylinder in the vicinity of a moving wall at upper-subcritical Reynolds number of $Re = 1.0 \times 10^5$. They successfully associated the methodologies developed by Bimbato et al. [11,21,22] focusing on the effect of higher relative roughness sizes, namely, $\varepsilon^*/d^* = 0.0045$ and 0.007, on flow dynamics of the cylinder at small-gap regime, which was identified in past investigation by Bimbato et al. [22] at $h^*/d^* < 0.20$. The small-gap regime is characterized by vortex shedding suppression. Their results shown an anticipation of the vortex shedding suppression when using $\varepsilon^*/d^* = 0.007$ at $h^*/d^* = 0.10$. Of importance, the Strouhal number completely vanished ($St = 0.0$) and it was observed the formation of two nearly parallel shear layers of opposite signals at the rear part of the cylinder. It is interesting to comment that Bimbato et al. [22] did not capture the complete interruption of vortex shedding, even at $h^*/d^* = 0.05$. This flow behavior will be investigated later in Section 3.2 to also contribute on discussion of the roughness model effect on it.

It is important to observe that Alcântara Pereira et al. [10] only computed the form component of the drag force. According to the discussions of Achenbach [9], the form (or pressure) component dominates the drag force on the smooth cylinder contributing more than 98% of the total drag force. On the other hand, the skin friction (or viscous) component of the drag force is responsible for the remaining 1–2%. In the literature, it is expected neither component of the drag force of the rough cylinder can be neglected. However, Achenbach [9] investigated the viscous drag force from a sand-roughened surface ($\varepsilon^*/d^* = 0.0011$ and 0.0045) and concluded that it contributed about 2–3% of the total drag force. That result showed a slight increase of the viscous drag force as compared as that of the smooth cylinder. Alcântara Pereira et al. [10] have reported differences above 10% between the smooth cylinder and other rough when integrating only static pressure, and, therefore, as discussed by them, the surface roughness effect has been captured by their numerical approach.

The main contribution of the present paper is shown that the methodology proposed by Alcântara Pereira et al. [10] also captures important changes in the flow dynamics of the slightly rough circular cylinder in the vicinity of a moving wall. Thus, the effect of small relative roughness size, namely, $\varepsilon^*/d^* = 0.001$, on flow dynamics around the cylinder is investigated and compared to smooth cylinder configurations. The chosen roughness size can be found, for example, in support columns of large offshore floating structures after a few years of operation. Put in other words, the main goal here is to report understanding of the influence of surface roughness for control of vortex shedding frequency reduction and wake destructive behavior behind the rough cylinder placed near a moving wall at upper-subcritical Reynolds number of $Re = 1.0 \times 10^5$. Our highlight results are for temporal history of aerodynamic loads, Strouhal number behavior and separation point prediction. The focus is to track the Strouhal number behavior until it vanishes. Overall, the results are found agree with the physics expected for this kind of vortical flow. The Section 4 will summarize the key findings of this study.

The rough circular cylinder aerodynamics has been reasonably reported in the literature [2,9,11,12,14,16,18], however, very little attention has been paid to the problem of the flow around the rough cylinder near a moving wall [10]. In recent past works [10,11], it has been reported that the effect of two-dimensional roughness model is much more sensitive than single turbulence modeling.

In general, numerical simulations of high Reynolds number flows around two-dimensional bluff bodies over predict aerodynamic forces behavior. However, the results are very important for applications of conservative designs in practical engineering problems, where higher integrated loads are computed, specially the drag force, in association with accurate vortex shedding frequencies. In the last years, our research group has made an effort to develop the in-house code for future extension to three-dimensional flows, and this research is integral part of the project.

2. Theory and Numerical Method

2.1. Physical Modeling

Figure 1 illustrates the smooth circular cylinder immersed in a semi-infinity fluid domain, Ω, with a free stream velocity, U^*, at infinity. The fluid is Newtonian with constant kinematic viscosity, ν. The flow is assumed to be unsteady, incompressible, and two-dimensional. The fluid domain can be identified by a surface $S = S_1 \cup S_2 \cup S_3$, being S_1 the cylinder surface, S_2 the moving wall surface and S_3 the far away boundary. The surface S is required to establish the boundary conditions of the physical problem. The location of the separation points of the flow for the top (open) and the bottom (gap) sides of the cylinder are defined by $\theta^+{}_{sep}$ and $\theta^-{}_{sep}$, respectively. The blockage effect is captured by reducing the gap-to-diameter ratio, h^*/d^*.

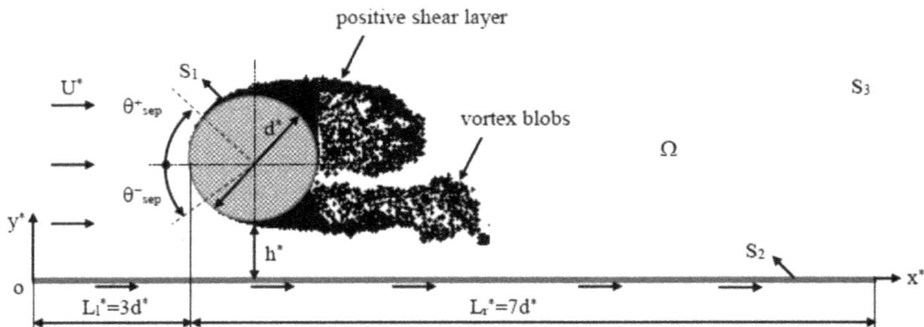

Figure 1. Schematic representation of the semi-infinity fluid domain.

To non-dimensionalize all the quantities of the problem, it is chosen the cylinder diameter, d^*, as scale length. The dimensionless time is defined by t^*U^*/d^*. In the general formulation, the symbol * is used to identify dimensional quantities.

The Reynolds number is defined in the following form:

$$Re = \frac{U^* d^*}{\nu} \tag{1}$$

According to Bearman [1], the presence of two shear layers of opposite signals is primarily responsible for vortex shedding, and the body merely modifies the mechanism by allowing feedback between the viscous wake and the shedding of circulation at the separation points (Figure 1). As consequence, the interaction between the two shear layers as a function of the free stream velocity, U^*, and the body diameter, d^*, is the key factor to define the rate at which vortical structures are cyclically detached at the rear part of the body. Thus, the dimensionless Strouhal number is defined as:

$$St \; = \; f^* \frac{d^*}{U^*} \tag{2}$$

being the frequency of vortex shedding, f^*, related with the scales d^* and U^*.

The Strouhal number for the smooth cylinder with no wall confinement has been experimentally measured in the literature and found close to $St \approx 0.19$ for an upper-subcritical Reynolds number of $Re = 1.0 \times 10^5$ [31]. In Equation (2), the quantity f^* is originally obtained from the temporal series of the lift force curve.

2.2. Introduction of Source Singularity Elements and Nascent Vortex Blobs

In the present numerical method, the smoothed cylinder and moving wall surfaces (Figure 1) are treated by an integral formulation of the potential component of the flow [32]. These surfaces are discretized and represented by flat panels with distribution of source singularity elements with constant density. Each flat panel has a center point, named pivotal point, where the impermeability boundary condition must be satisfied. The impermeability condition is imposed on each pivotal point and it establishes equality between the normal velocity component of a fluid particle and the normal velocity component of each pivotal point. Of numerical importance is the fact that the impermeability condition and the mass conservation of the problem are guaranteed by the source elements generation in each temporal step.

In addition to this, the discrete vortex method engages in to discretize spatially the vorticity field using an instantaneous vortex blobs collection, as illustrated in Figure 1. The vortex blobs are represented by Lamb vortex elements, each one presenting a distribution of vorticity, ς_{σ_0}, (commonly called the cut-off function), a circulation strength, Γ, a core size, σ_0, and a spatial position, x [33,34]. The no-slip boundary condition is imposed on each pivotal point and it establishes equality between the tangential velocity component of a fluid particle and the tangential velocity component of each pivotal point. Of numerical importance is the fact that the no-slip condition and the global circulation conservation of the problem are guaranteed by the vortex blobs generation in each temporal step. The moving wall effect dispenses this boundary condition on the ground plane, being the vortex blobs only generated on the cylinder surface [21,22,25]. Figure 2 illustrates as a vortex blob is introduced into the fluid domain. In Lagrangian manner, the vorticity generated from a flat panel stay concentrated inside blob vortex core, σ_0.

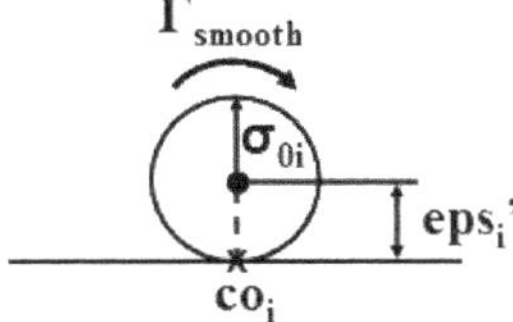

Figure 2. Shedding of a vortex blob during one time stepping (co_i is the pivotal point and eps_i' defines the "smooth shedding point" location of the vortex blob).

To satisfy the boundary conditions above mentioned are necessary two different kinds of system of linear algebraic equations, which are solved iteratively using the method of least squares. The simultaneous generation of sources elements and vortex blobs can also be coupled with the roughness model, when activated; details of the roughness model used in simulations of interest will be presented in Section 2.4.

2.3. Discrete Vortex Method with LES Modeling

The motion of each vortex blob is governed by the vorticity transport equation, which is obtained by taking the curl of the Navier–Stokes equations [33]. Chorin [35] proposed an algorithm that splits the vorticity transport equation to separately solve the advection and diffusion problems. Alcântara Pereira et al. [36] originally presented the solution of the two problems including LES modeling into the two-dimensional discrete vortex method, such as, respectively:

$$\frac{D\overline{\omega}}{Dt} = \frac{\partial\overline{\omega}}{\partial t} + (\overline{u} \cdot \nabla)\overline{\omega} = 0 \tag{3}$$

$$\frac{\partial\overline{\omega}}{\partial t} = \left(\frac{1}{Re} + v_t^*\right)\nabla^2\overline{\omega} \tag{4}$$

where

$$v_t^* = \frac{v_t}{U^*d^*} \tag{5}$$

represents the local eddy viscosity coefficient and $\overline{\omega} = \nabla \times \overline{u}$ defines the vorticity scalar field.

The vorticity generated from each flat panel in a time stepping is regarded "free" to undergo advection and diffusion process satisfying Equations (3) and (4), respectively. The vortex blobs transport by advection (Equation (3) is computed through the following expression:

$$x_i^{t+\Delta t} = x_i^t + u(x_i)^{t+\alpha\Delta t}\Delta t = x_i^t + \left[(K*\overline{\omega})(x) + \overline{ub_i^n}(x_i) + 1\right]^{t+\alpha\Delta t}\Delta t \tag{6}$$

where $u(x_i)^{t+\alpha\,\Delta t}$ represents the velocity vector of the filtered field, and α is the temporal integrating parameter, such that $0 \le \alpha \le 1$, where $\alpha = 0$ and $\alpha = 1$ define an explicit and an implicit Euler scheme, respectively. In this work, is adopted an explicit Euler scheme.

In Equation (6), the velocity vector is computed at the point occupied by the *i*th vortex blob according to the Biot–Savart law (vortex–vortex interaction), panel method (vortex–panel interaction) and free stream velocity (vortex–mainstream interaction) contributions, respectively. It is remarkable that the Lagrangian manner [37] dispenses the need to explicitly treat advective derivatives into the Equation (3).

Equation (4) is solved using to the random walk method [35], where is imposed a displacement for each vortex blob in the following form [36]:

$$\varsigma_i(t) = \sqrt{\frac{4\Delta t}{Re_{c_i}}ln\left(\frac{1}{p}\right)}\left[\cos(2\pi Q)_x + \sin(2\pi Q)_y\right] \tag{7}$$

being p and Q random numbers generated between 0.0 and 1.0, and Re_c the local Reynolds number modified by the local eddy viscosity coefficient, such as [36]:

$$Re_{c_i}(t) = \frac{U^*d^*}{v + v_{t_i}(t)} \tag{8}$$

In Equation (7), p and Q define two random displacements with mean equal to zero and a variance given by twice the product of the kinematic viscosity and the time. The vorticity diffusion process through the random displacements of vortex blobs simulates the viscosity effect.

In the present formulation, the local turbulence effect is simulated during the vorticity diffusion process. Therefore, the local eddy viscosity coefficient computation is necessary to include the effect of the small scales through the concept of differences of velocity between vortex blobs [36]. The support to the turbulence modeling success is that each vortex blob needs to move with the local fluid velocity in a Lagrangian manner to simulate the vorticity advection process, and the velocity induced at each vortex blob (see $u(x_i)^{t+\alpha\,\Delta t}$ in Equation (6)) is calculated before the vorticity diffusion process.

The average velocity differences are required to calculate the second-order velocity structure function of the filtered field [38] for each vortex blob that constructs the viscous wake. Alcântara Pereira et al. [36] proposed an adaption to compute the second-order velocity structure function in two-dimensions, such as:

$$\overline{F}_{2_j} = \frac{1}{N}\sum_{k=1}^{N}\|\overline{u}_{t_j}(x_j) - \overline{u}_{t_j}(x_j + r_k)\|_k^2\left(\frac{\sigma_{0_j}}{r_k}\right)^{2/3} \tag{9}$$

where j defines the position of jth vortex blob, $\overline{u}_t$ defines the velocity filtered field computed at each vortex blob, N characterizes a special group of vortex blobs inside a circular crown idealized around the jth vortex blob under analysis, and r_k measures the distance between the jth and kth vortex blobs, the latter necessarily belonging to that special group (for more details, please see Bimbato et al. [11]).

After the solution of Equation (9), the local eddy viscosity coefficient computation is obtained through the following formula [38]:

$$v_{t_i}(t) = 0.105C_k^{-3/2}\sigma_{0_k}\sqrt{\overline{F}_{2_i}(t)} \tag{10}$$

being $C_k = 1.4$ the Kolmogorov constant.

It is important comment that use of two-dimensional LES-based turbulence modeling is necessary to stabilize the numerical solution of the problem; furthermore, it also provides a basis for a future three-dimensional turbulence modeling. Bimbato et al. [21,22] validated the LES modeling used in this work.

As already mentioned in Section 2.2, the smoothed cylinder surface is represented by source flat panels [32], being that each one also produces one vortex blob of strength Γ_i at every time stepping. Over a period of time, the viscous boundary layer develops to take on the form of a cluster of several thousand vortex blobs, as can be seen from the computer output shown in Figure 3. The separation that occurs on the cylinder surface originates an unsteady flow with the presence of Von Kármán large-scale vortices downstream the body. The proximity of a moving wall (blockage effect) will certainly interfere in any way with that vortex formation regime.

Figure 3. Starting flow around the smooth circular cylinder with no wall confinement.

It is remarkable that with the Lagrangian tracking of vortex blobs, one need not take into account the far away boundary conditions, i.e., at S_3 in Figure 1. In addition, the computations are only concentrated on regions containing significant vorticity, which are the regions of high activities of the flow. The Lagrangian formulation of the discrete vortex method indeed dispenses a grid for the spatial discretization of the interest domain. Thus, numerical instabilities associated to high Reynolds number flows do not require special care in contrast to the Eulerian schemes. In order to take care of both advection and diffusion of the vorticity, one makes use of an advection-diffusion splitting algorithm [35]; according to it the advection of each vortex blob is carried out independently of the diffusion (Equations (6) and (7)). In other cases, when vortex blobs migrate to the interior of a solid surface, they are reflected from their paths.

2.4. The Roughness Model

The surface roughness effect is associated with the vortex blobs creation process, such that the circulation strength, Γ_{smooth}, of each nascent vortex blob is increased by the amount of $\Delta\Gamma$. The value of $\Delta\Gamma$ is defined by the turbulent activity around a "smooth shedding point" (Figure 2) and accounting by changing the vorticity of the nascent vortex blob. Adjacent to each flat panel, used to represent the cylinder surface, is simulated an inertial effect promoted by body roughness.

The key idea of the roughness model capturing changes in the vortex shedding frequency (Strouhal number) is sketched in Figure 4a, where co_i represents a pivotal point and $eps_i(t)$ defines the location of a vortex blob, after its vorticity to be changed by surface roughness effect. The numerical effect of the roughness model is to change the core size, σ_0, of each nascent vortex blob, using the following equation [21]:

$$\sigma_{0c_k}(t) = 1.41421 \sqrt{\frac{\Delta t}{Re}\left(1 + \frac{v_{t_i}(t)}{v}\right)} \tag{11}$$

being Δt the temporal step estimated from velocity scale of the flow and Re the Reynolds number of the flow, as defined in Equation (1).

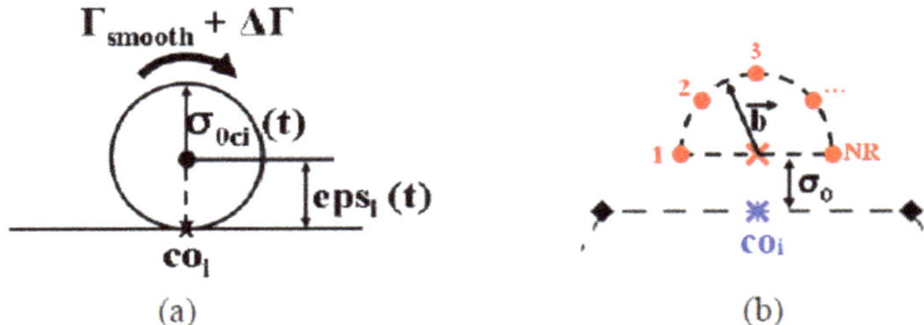

Figure 4. (**a**,**b**) Shedding of a vortex blob during one time stepping (co_i is the pivotal point and $eps_i(t)$ defines the "rough shedding point" location of the vortex blob).

In Equation (11), the local eddy viscosity coefficient computation newly depends on the solution of Equation (10). Therefore, the average velocity differences, required to calculate the second-order velocity structure function of the filtered field and to simulate surface roughness effect, are evaluated between the center of a semicircle with radius of $b = 2\varepsilon - \sigma_0$ and rough points, both placed near to each pivotal point, as follow [11]:

$$\bar{F}_{2_i}(t) = \frac{1}{NR} \sum_{w=1}^{NR} \|\bar{u}_{t_i}(x_i, t) - \bar{u}_{t_w}(x_i + b, t)\|^2_w (1 + \varepsilon) \tag{12}$$

where u_t is the instantaneous velocity filtered field, NR defines twenty-one rough points adjacent to each flat panel (Figure 4b), and $(1 + \varepsilon)$ characterizes the kinetic energy gain associated with the average roughness effect, $\varepsilon = \varepsilon^*/d^*$ (for more details, please see Bimbato et al. [11]).

2.5. Aerodynamic Loads

As previously commented in Section 2.3, the only governing equation in the discrete vortex method is the vorticity transport equation of the filtered field. It should be remarked that the pressure is missing in the formulation, because the pressure term is eliminated when is applied the curl in the

Navier–Stokes equations. The pressure filtered field is then recuperated by taking the divergence operator to the Navier–Stokes equations. The procedure starts with the stagnation pressure definition, such as:

$$\overline{Y}^* = \frac{\overline{p}^*}{\rho} + \frac{\overline{u}^{*2}}{2}, \quad \overline{u}^* = \left|\overline{u}^*\right| \tag{13}$$

where $\overline{p}^*$ defines the static pressure, ρ is the fluid density and $\overline{u}^*$ represents the velocity.

Thus, a Poisson equation is derived, and the static pressure is obtained using an integral formulation, such as [39,40]:

$$H\overline{Y}_p - \int_{S_1+S_2} \overline{Y}\nabla\Xi_p \cdot e_n dS = \iint_\Omega \nabla\Xi_p \cdot (\overline{u}\times\overline{\omega})d\Omega - \frac{1}{Re}\int_{S_1} \left(\nabla\Xi_p \times\overline{\omega}\right)\cdot e_n dS \tag{14}$$

where p is the any point to compute pressure, $H = 1$ for points of the fluid domain, $H = 0.5$ for pivotal points, Ξ represents the fundamental solution of Laplace equation, and e_n defines the unit vector normal to each solid boundary.

Finally, the drag and lift coefficients are calculated as, respectively [21]:

$$C_D = \sum_{p=1}^{NP} 2\left(\overline{p}_p - p_\infty\right)\Delta S_p \mathrm{sen}\beta_p = \sum_{p=1}^{NP} C_{P_p}\Delta S_p \mathrm{sen}\beta_p \tag{15}$$

$$C_L = -\sum_{p=1}^{NP} 2\left(\overline{p}_p - p_\infty\right)\Delta S_p \cos\beta_p = -\sum_{p=1}^{NP} C_{P_p}\Delta S_p \cos\beta_p \tag{16}$$

where NP represents the total number of flat panels, p_∞ represents the reference pressure at S_3 (Figure 1), ΔS_p defines a flat panel length, and β_p defines a flat panel angle.

2.6. Computational Sequence for Solution of the Navier–Stokes Equations

The numerical method described above is implemented to run sequentially according to the followings steps: (i) simultaneous generation of source panels and nascent vortex blobs (including the roughness model); (ii) calculation of the velocity vector at the point occupied by each vortex blob; (iii) calculation of surface pressure distribution and hence drag and lift on the cylinder; (iv) advection of the vortex blobs; (v) diffusion of the vortex blobs (including the LES modeling); (vi) reflection of the vortex blobs that migrate into the cylinder or ground plane; (vii) calculation of the velocity induced by vortex cloud at pivotal points; and (viii) advance by time Δt.

3. Results and Discussion

3.1. Simulation Setup

The chosen upper-subcritical Reynolds number for all computations was $Re = 1.0 \times 10^5$, which allows comparisons with experimental results presented by Nishino [25], when possible. Test cases were previously performed for the smooth circular cylinder aiming to find suitable values for the following parameters [21,22]: dimensionless time, $t = 75$; dimensionless time step, $\Delta t = 0.05$; number of nascent vortex blobs during each time step, $M = 300$; and initial displacement of each nascent vortex blob adjacent to the flat panel, $eps_i' = \sigma_{0i} = 0.001$, as illustrated in Figure 2.

The dimensionless time step was estimated according to $\Delta t = 2\pi K/N$, being $0 < K < 1$ and $N = 300$ [21,22] and it depends on the accuracy of the explicit Euler scheme; in this procedure N represents the number of flat panels adopted to discretize the cylinder surface [32]. All obtained values were computed in the interval $37.50 \le t \le 75.00$ to compute time-averaged results. The chosen dimensionless time step was found suitable to compute aerodynamic loads with accuracy reducing the final time of the simulations.

In Figure 1, the wall plane length was fixed as $L^*/d^* = 10$ and discretized using 950 flat panels. A horizontal distance of $L_l^*/d^* = 3$ allows to identify the cylinder front stagnation point, where the reference starts from the origin of inertial frame of reference placed at $(x^*/d^*; y^*/d^*) = (0.0; 0.0)$. That horizontal distance was previously investigated, being the length of $L_l^*/d^* = 3$ found suitable to capture the wall confinement (blockage) effect [21,22]. The blockage effect was captured when the gap-to-diameter ratio, h^*/d^*, was reduced from 0.80 to 0.05 (Tables 1 and 2). The relative roughness size of $\varepsilon^*/d^* = 0.001$ was chosen for the test cases, when the roughness model was activated. All the numerical results with no roughness model were identified by $\varepsilon^*/d^* = 0.000$ in Section 3.2.

Table 1. Experimental and numerical data of aerodynamic force coefficients for the circular cylinder in the vicinity of a moving wall ($Re = 1.0 \times 10^5$).

Gap-to-Diameter Ratio	Nishino [25] with No End Plates		Nishino [25] Using End Plates ($y_e^*/d^* = 0.00$)		Nishino [25] Using End Plates ($y_e^*/d^* = 0.40$)		Present Simulation ($\varepsilon^*/d^* = 0.000$)		Present Simulation ($\varepsilon^*/d^* = 0.001$)	
h^*/d^*	C_D	C_L	C_D	C_L	C_D	C_L	C_D	C_L	C_D	C_L
0.8	0.899	0.014	1.293	−0.02	1.385	0.024	1.38	0.092	1.284	0.04
0.6	0.92	0.039	1.302	0.001	1.373	0.038	1.408	0.105	1.437	0.103
0.5	0.924	0.045	1.282	0.034	1.323	0.09	1.474	0.104	1.459	0.104
0.45	0.926	0.06	1.242	0.054	1.311	0.102	1.443	0.108	1.392	0.105
0.4	0.922	0.074	1.145	0.084	-	-	1.433	0.058	1.454	−0.003
0.35	0.931	0.092	0.929	0.078	-	-	1.44	0.029	1.393	−0.002
0.3	0.93	0.117	0.941	0.111	-	-	1.426	−0.031	1.473	−0.026
0.25	0.933	0.144	0.951	0.154	-	-	1.455	−0.016	1.339	−0.041
0.2	0.939	0.177	0.954	0.188	-	-	1.435	0.004	1.278	−0.015
0.15	0.952	0.231	0.957	0.247	-	-	1.281	0.056	1.151	0.068
0.1	0.958	0.308	0.953	0.306	-	-	1.081	0.411	0.897	0.301
0.05	0.965	0.429	0.941	0.477	-	-	0.877	0.531	0.559	0.313

Table 2. Summary of data for the Strouhal number and position of separation for the circular cylinder in the vicinity of a moving wall ($Re = 1.0 \times 10^5$).

Gap-to-Diameter Ratio	Present Simulation ($\varepsilon^*/d^* = 0.000$)	Present Simulation ($\varepsilon^*/d^* = 0.001$)	Present Simulation ($\varepsilon^*/d^* = 0.000$)		Present Simulation ($\varepsilon^*/d^* = 0.001$)		Nishino [25] with No End Plates	
h^*/d^*	St	St	θ^+_{sep}	θ^-_{sep}	θ^+_{sep}	θ^-_{sep}	θ^+_{sep}	θ^-_{sep}
0.8	0.213	0.202	83°	88°	85°	90°	78°	82°
0.6	0.204	0.212	81°	88°	82°	95°	-	-
0.5	0.204	0.216	79°	91°	85°	96°	-	-
0.45	0.205	0.208	80°	92°	84°	99°	-	-
0.4	0.199	0.212	79°	93°	85°	97°	72°	92°
0.35	0.199	0.201	79°	95°	86°	98°	-	-
0.3	0.198	0.199	78°	94°	82°	99°	-	-
0.25	0.195	0.189	76°	95°	83°	100°	-	-
0.2	0.174	0.166	75°	97°	83°	104°	64°	91°
0.15	0.143	0.129	72°	99°	79°	104°	-	-
0.1	0.113	0.087	69°	100°	82°	100°	-	-
0.05	0.107	0.0	70°	103°	76°	102°	-	-

3.2. Circular Cylinder in the Vicinity of a Moving Wall

This section presents simultaneous measurements of integrated aerodynamic loads and surface pressure distributions for the circular cylinder, which are essentials to support all bellow discussions. The main objective is to report that the roughness model associated with blockage effect captures the full interruption of vortex shedding from the slightly rough cylinder surface placed closer to a moving wall. Thus, the results of drag force reduction, positive lift force, Strouhal number behavior, and location of the separations points of the flow around the cylinder will successfully support the analyses in a very good physical sense.

Table 1 presents experimental and numerical results of the aerodynamic force coefficients for the cylinder at upper-subcritical Reynolds number flows of $Re = 1.0 \times 10^5$. The experimental results are for the smooth cylinder case at different gap-to-diameter ratios, $h*/d*$, being the uncertainties in the drag, C_D, and lift, C_L, coefficients of ±0.016 and ±0.011, respectively, with 95% confidence [25]. The goal is to compare them to our numerical results, also presented in Table 1.

The experimental results presented in Table 1 and identified as "using end plates" were obtained for the ratios of $y_e*/d* = 0.00$ and 0.40. In the experiments of Nishino [25], the length y_e* was defined as the distance from the edge of the cylinder to the bottom border of the end plate. In the experimental study, the bottom border of the end plate was placed below the cylinder, being the edge of the cylinder its bottom side ($\theta^+_{sep} = 270°$ in Figure 1).

With reference to Table 1, the experiments of Nishino [25] revealed that the drag force increases with the use of end plates for $h*/d* \geq 0.45$, i.e., the flow becomes more two-dimensional. The use of end plates with $y_e*/d* = 0.40$ revealed the flow closest to a two-dimensional pattern. The use of a pair of end plates, especially at high Reynolds numbers flows, was justified by Nishino [25] since the effect of the end condition of the cylinder cannot usually be large enough in practical investigations. As comparison, the numerical results of drag force for the smooth cylinder ($\varepsilon*/d* = 0.000$) present a very good agreement with the end plates configuration at $y_e*/d* = 0.40$, being the difference of the drag coefficient between them around 10%. When the small-gap regime is identified for $h*/d* < 0.20$, the drag force significantly reduces, and this behavior occurs because of the surface roughness effect. Lei et al. [23] observed that the critical drag behavior cannot be accurately determined, since experimental and numerical investigations are carried out using discrete gap-to-diameter ratios, and the vortex shedding suppression manifests as ratio $h*/d*$ is gradually reduced. In general, the numerical results show that the lift force for the rough cylinder is slightly lower as compared as the smooth cylinder; that behavior agrees with observations [2].

Table 2 summarizes results of the Strouhal number, St, and separation point prediction, θ^+_{sep} and θ^-_{sep} (Figure 1), for the same study cases shown in Table 1. There are no experimental data available of Strouhal number for the flow around smooth and rough cylinders near a moving wall. The experimental results for position of separation of the flow past the cylinder with no end plates [25] were included for comparison purposes. Nishino [25] did not report experimental results of the separation points prediction for the configuration of the cylinder using end plates at $y_e*/d* = 0.40$.

As illustration, the temporal history of the drag and lift coefficients of the smooth cylinder and other rough, both at $h*/d* = 0.50$ and 0.05, is shown in Figure 5.

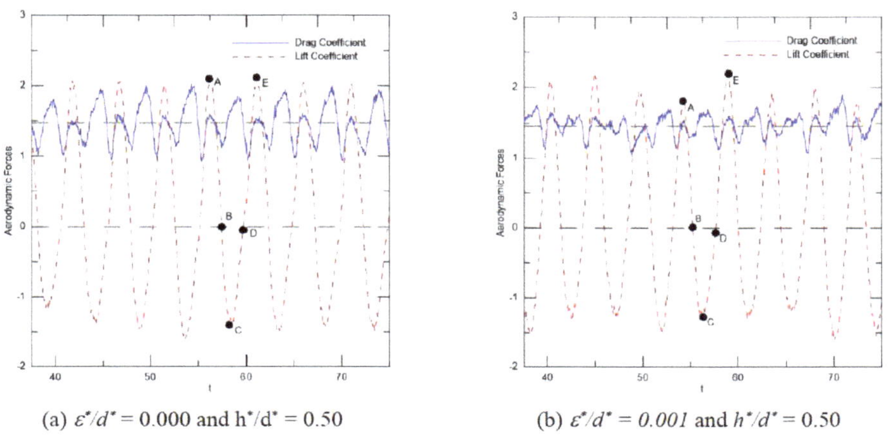

(a) $\varepsilon*/d* = 0.000$ and $h*/d* = 0.50$ (b) $\varepsilon*/d* = 0.001$ and $h*/d* = 0.50$

Figure 5. *Cont.*

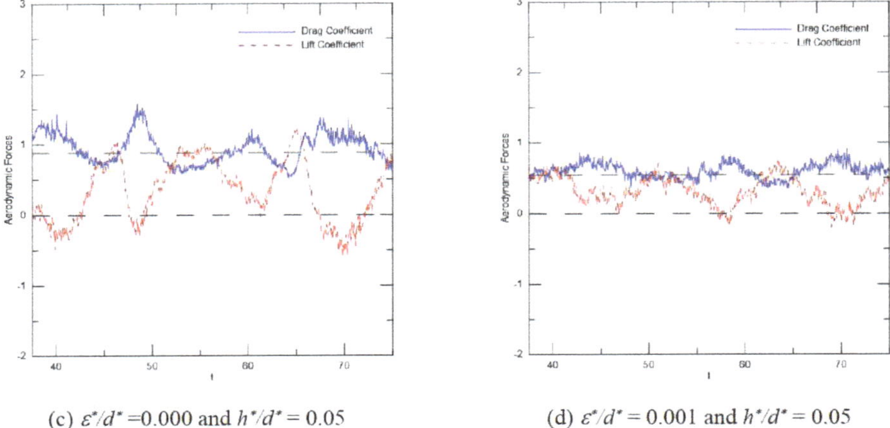

(c) $\varepsilon^*/d^* = 0.000$ and $h^*/d^* = 0.05$ (d) $\varepsilon^*/d^* = 0.001$ and $h^*/d^* = 0.05$

Figure 5. Temporal history of the drag and lift coefficients for the circular cylinder in the vicinity of a moving wall ($Re = 1.0 \times 10^5$).

Figure 6a,b present instantaneous pressure coefficient, C_P, distributions related to instants represented by points A–E, marked in Figure 5a,b, for the cylinder at $h^*/d^* = 0.50$. In Figure 6a,b, the separation angle on upper (open) side of the cylinder is identified as $\theta \equiv \theta^+_{sep}$.

In Figure 5a, the instant represented by point A characterizes a maximum value of the lift coefficient, where a clockwise vortical structure detaches from the smooth cylinder upper surface and moves toward the near wake (Figure 7a). That vortical structure grows and starts to attract the shear layer of opposite signal (Figure 7b). As consequence, a low pressure region is created at the cylinder upper surface (Figure 6a).

The instant represented by point C in Figure 5a indicates a minimum value of the lift coefficient, where a counter-clockwise vortical structure detaches from the smooth cylinder lower surface and moves toward the near wake. That vortical structure can be visualized in Figure 7c and it creates a low pressure region at the cylinder lower surface (Figure 6a). The same counter-clockwise vortical structure grows and starts to attract the shear layer of opposite signal (Figure 7d), the latter is feeding the clockwise vortical structure causing its detachment. It is important to observe that a new upper vortical structure will be born on the upper surface and will start to grow attracting the lower shear layer; the latter also will feed the lower vortical structure causing its detachment. The complete incorporation to the near wake of the lower and upper vortical structures is revealed at instants identified by points B and D in Figure 7b,d, respectively. In Figure 6a, the instant defined by point E represents the same event previously described for the point A.

The mechanism above reported is cyclic repeating alternatively on the upper and lower sides of the smooth cylinder surface. Thus, an unsteady flow with the presence of von Kármán-type vortices takes place downstream of the cylinder (Figure 8a). That phenomenon agrees to the classical vortex shedding mechanism of the smooth cylinder with no ground effect [41]. In Figure 8a, the viscous wake downstream takes the form of "mushroom-type" vortical structures, being that the blockage effect will destroy them far from the cylinder. For the large-gap regime, antisymmetrical perturbations are captured from the near wake region and they are felt near the cylinder surface. These perturbations intrinsically relate to the Von Kármán large-scale vortices formation mode.

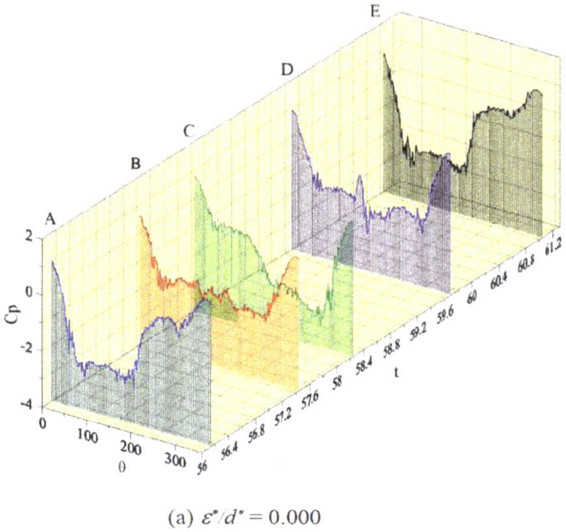

(a) $\varepsilon^*/d^* = 0.000$

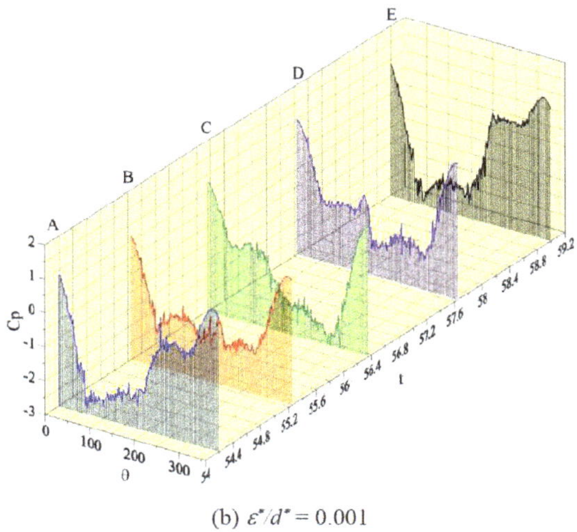

(b) $\varepsilon^*/d^* = 0.001$

Figure 6. Instantaneous pressure distributions for the circular cylinder in the vicinity of a moving wall at $h^*/d^* = 0.50$ ($Re = 1.0 \times 10^5$).

Gerrard [41] stated that a vortical structure shedding from bluff body surface continues to grow (see, e.g., the upper vortical structure in Figure 7b), being fed by circulation originated of the connected shear layer and that, when it is strong enough, draws the opposing shear layer across the near wake. In this mechanism, the approximation of the shear layers of opposite signals is able to cut off further supply of circulation to the growing vortical structure, the latter, then, sheds and moves downstream the cylinder.

In this work, the mechanism of vortical structures formation at the rear part of the cylinder with no wall confinement [41] has also been identified for the smooth cylinder at $h^*/d^* = 0.50$ (Figure 7a–d).

Therefore, it can be concluded that the Venturi effect really redraws the lower vortical structure shed downstream. Figure 7c gives us an idea of the Venturi effect acting on a nascent vortical structure and deforming it.

The Venturi effect also contributes creating two different highest peaks for the drag coefficient curve, which is synchronized with the lift coefficient curve (Figure 5a). The explication for this interesting aerodynamic force behavior is that while the upper vortical structure finds total freedom to grow at the rear part of the cylinder until to be incorporated by the viscous wake (Figure 7b), leading to a bigger value in the drag coefficient (see approximately the drag coefficient value by projecting point C in Figure 5a), the developing of the lower vortical structure is affected by the Venturi effect (Figure 7c). This second event reflects the smaller peak in the drag curve (see approximately the drag coefficient value by projecting point D in Figure 5a).

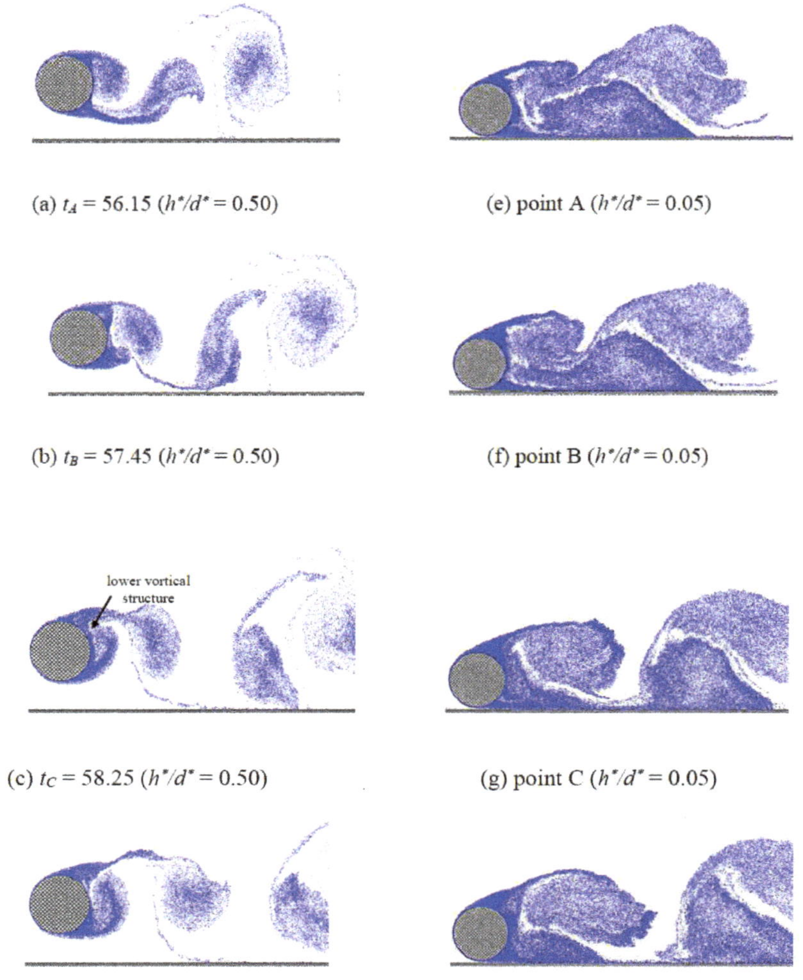

(a) $t_A = 56.15$ ($h^*/d^* = 0.50$)

(e) point A ($h^*/d^* = 0.05$)

(b) $t_B = 57.45$ ($h^*/d^* = 0.50$)

(f) point B ($h^*/d^* = 0.05$)

(c) $t_C = 58.25$ ($h^*/d^* = 0.50$)

(g) point C ($h^*/d^* = 0.05$)

Figure 7. Near wake patterns for the circular cylinder in the vicinity of a moving wall ($\varepsilon^*/d^* = 0.000$; $Re = 1.0 \times 10^5$).

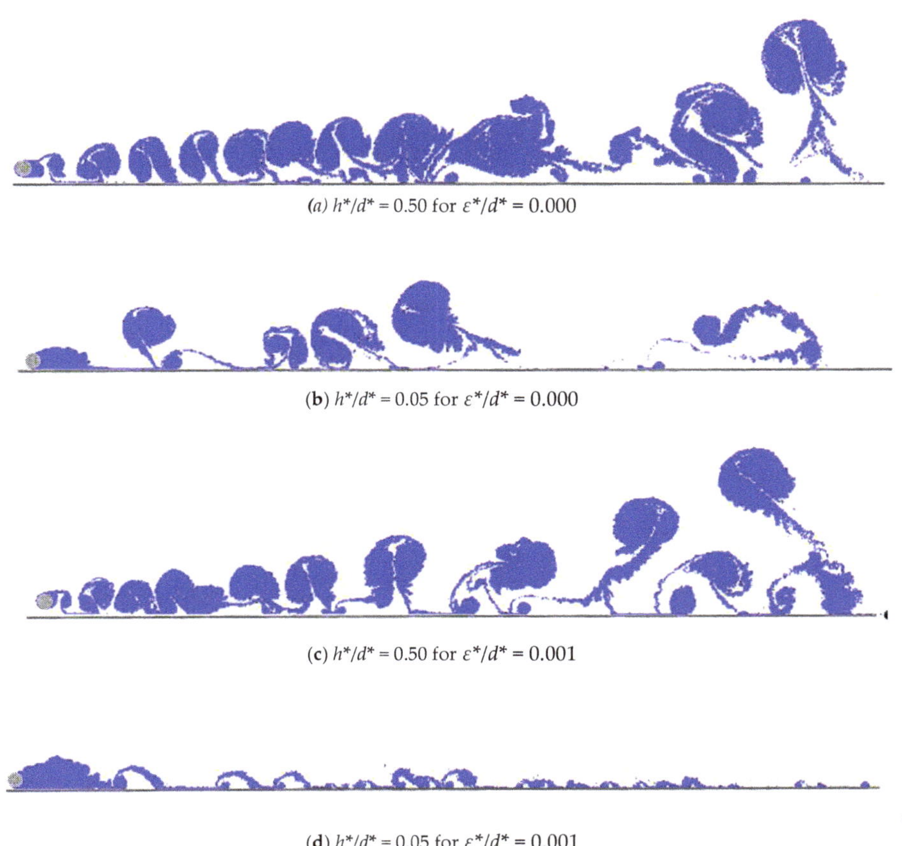

(a) $h^*/d^* = 0.50$ for $\varepsilon^*/d^* = 0.000$

(**b**) $h^*/d^* = 0.05$ for $\varepsilon^*/d^* = 0.000$

(**c**) $h^*/d^* = 0.50$ for $\varepsilon^*/d^* = 0.001$

(**d**) $h^*/d^* = 0.05$ for $\varepsilon^*/d^* = 0.001$

Figure 8. Final positions of clusters of vortex blobs for the circular cylinder in the vicinity of a moving wall at $t = 75$ ($Re = 1.0 \times 10^5$).

The study case for the smooth cylinder at $h^*/d^* = 0.50$ forecasts the occurrence of a (mean) stagnation point near the frontal part of the cylinder, where, at that position, the pressure coefficient is around 1.0 at $\theta \approx 0.6°$ (Figure 6a). It is important to observe that the front stagnation point of the cylinder with no wall confinement is located at $\theta_{stag} = 0.6°$ because the discretization of the cylinder surface using the panel method [32] initiates at $\theta = 0°$, with the first pivotal point located at $\theta = 0.6°$. The upper (open) separation angular position is identified at about $\theta^+_{sep} \approx 79°$ (Table 2). The experimental result for the smooth cylinder with no wall confinement at $Re = 1.0 \times 10^5$ is predicted around $\theta \approx 82°$ [31]. The separation angle at bottom (gap) side is about $\theta^-_{sep} \approx 91°$ (Table 2). This result is consistent with the expected physics for the problem because the wall confinement (blockage) effect really changes downstream the separation angle at cylinder bottom side. The experimental result of Nishino [25] for the cylinder with no end plates at $h^*/d^* = 0.40$ is predicted to occur about $\theta^-_{sep} \approx 92°$ (Table 2). It is important to observe that a lower rear pressure (Figure 6a), identified for the cylinder in ground effect, reflects a higher value for the drag coefficient ($C_D \approx 1.474$ in Table 1) as compared to the experimental value reported by Blevins [31], that is $C_D \approx 1.2$, with 10% uncertainty. As additional information, a numerical result available in the literature [10] for smooth cylinder with no wall confinement is about $C_D \approx 1.198$.

The numerical result of the mean lift coefficient for the smooth cylinder at $h^*/d^* = 0.50$ is predicted around $C_L \approx 0.104$ (Table 1). The appearance of lift force pointing away from the moving wall is because of the viscosity effect, which definitively contributes to move the cylinder front stagnation point downstream and, as consequence, it is created an additional positive circulation with increasing of the lift force. This change on the front stagnation point is captured through the temporal history of points A–E in Figure 6a, in which a slight change can be identified (i.e., C_p is not more equal to 1.0 at $\theta = 0.6°$). For the smooth cylinder with no wall confinement, the lift force oscillates around mean value of zero [10].

The numerical result of the Strouhal number for the cylinder at $h^*/d^* = 0.50$ seems insensitive to blockage effect, being predicted around $St \approx 0.204$ (Table 2). This conclusion is supported by the experimental value available in the literature for the smooth cylinder with no wall confinement at upper-subcritical flow regime of $Re = 1.0 \times 10^5$, that is around $St \approx 0.19$ [31], and also with 10% uncertainty. In accordance, the vortices formation mechanism [41] is not delayed, which justifies the Strouhal number value does not change. The latter conclusion is substantiated by defining the period $t_E - t_A$ corresponding to the detachment of a pair of vortical structures from cylinder surface and connected to each other by a vortex sheet, which rotate in opposite directions until they be completely incorporated into the viscous wake. The period computed is of $t_E - t_A = 61.15 - 56.15 = 5.0$ for the smooth cylinder at $h^*/d^* = 0.50$ (Figure 6a) and of $t_E - t_A = 4.7$ for the smooth cylinder with no wall confinement [10], being the difference between them around 6.0%.

The saturation state of a typical numerical simulation at dimensionless time of $t = 75$ can be demonstrated through of the difference around 2%between the Strouhal number, obtained from the inverse of the period $t_E - t_A \approx 0.20$ (Figure 6a), and other one of $St \approx 0.204$, which was computed from a Fast Fourier Transformation of the lift curve between $37.5 \leq t \leq 75$ (Figure 5a).

On the other hand, the temporal history of the drag and lift coefficients for the smooth cylinder at $h^*/d^* = 0.05$ can be seen in Figure 5c. The drag reduction is about 40.5% as compared as the smooth cylinder at $h^*/d^* = 0.50$ (Table 1). Figure 7e–h sketch the near wake pattern for the smooth cylinder at $h^*/d^* = 0.05$, being the instants defined by respective points A–D in Figure 9a. The Strouhal number reduces to $St \approx 0.107$ (Table 2), which characterizes intermittency on the von Kármán large-scale vortex formation mode (Figure 8b).

It is of great importance for the present work that the moving wall control demonstrates the efficiency of the roughness model to reduce the drag force of the rough cylinder, when the passive control technique of vortex shedding is activated, for the chosen relative roughness size of $\varepsilon^*/d^* = 0.001$ (Figure 5b,d can be compared). The hybrid control technique is therefore able to reduce the drag force of the smooth cylinder at $h^*/d^* = 0.50$ around 62.1% as compared as the rough cylinder at $h^*/d^* = 0.05$ (from $C_D \approx 1.474$ to 0.559 in Table 1). It is interesting to comment that Alcântara Pereira et al. [10] also identified a strong reduction on the drag force when using the relative roughness size of $\varepsilon^*/d^* = 0.007$ at $h^*/d^* < 0.20$ (small-gap regime). In their numerical experiment, the drag force of the rougher cylinder at $h^*/d^* = 0.05$ reduced around 60.2% as compared as the smooth cylinder near a moving wall at $h^*/d^* = 0.50$. In both numerical studies, the higher drag reduction is because of the passive control technique promoting the full interruption on the von Kármán large-scale vortices formation mode.

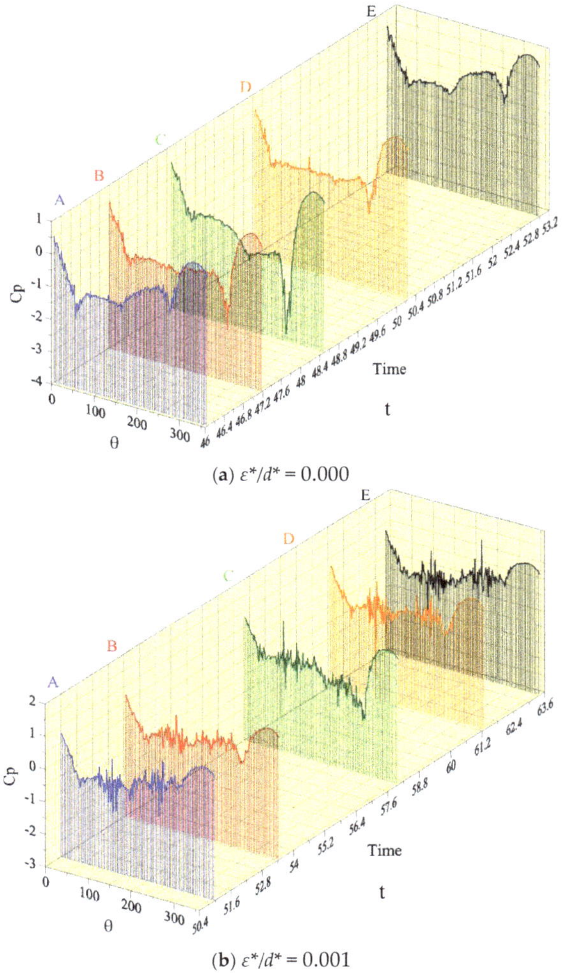

(**a**) $\varepsilon^*/d^* = 0.000$

(**b**) $\varepsilon^*/d^* = 0.001$

Figure 9. Instantaneous pressure distributions for the circular cylinder in the vicinity of a moving wall at $h^*/d^* = 0.05$; $Re = 1.0 \times 10^5$.

The rough cylinder at $h^*/d^* = 0.05$ configuration is strongly sensitive to interfere to the Venturi effect, which controls the smooth cylinder aerodynamics when submitted to ground effect. The surface roughness effect also participates to control the flow dynamics of the cylinder. The lift coefficient reduces of $C_L \approx 0.531$ (smooth case) for $C_L \approx 0.313$ (rough case), a reduction of 41.1% (Table 1).

Figure 5b shows that the surface roughness effect interferes on the orderly behavior of the big and small peaks previously identified in the drag curve for the smooth cylinder at $h^*/d^* = 0.50$ (Figure 5a). Now, in Figure 5b, there are no more single big and small peaks as the dimensionless time runs from $t = 40$ on, approximately. In Figure 5b, it is until difficult to identify the big and small peaks in the drag curve because of noise increasing originated from roughness effect. The period $t_E - t_A$ is of 5.0 in Figure 6b, and of $t_E - t_A = 4.7$ for the cylinder with no wall confinement [10]. Once again, the difference between them is around 6.0%, and the suppression of vortex shedding cannot be promoted when using $\varepsilon^*/d^* = 0.001$ at $h^*/d^* = 0.50$ (Figure 8c).

Figure 9b presents instants randomly chosen and represented by points A–E for instantaneous pressure distributions of the rough cylinder at $h^*/d^* = 0.05$. The simulation also predict the appearing of a (mean) stagnation point near the cylinder frontal part, where, at that position, the pressure coefficient, C_p, is not more 1.0 at $\theta = 0.6°$ (Figure 9b). In Figure 5d, is more difficult to identify points A–E as compared as the Figure 5a.

The Strouhal number completely vanishes, $St = 0.0$, for the cylinder with roughness model at $h^*/d^* = 0.05$ (Table 2). Bimbato et al. [22] also identified a decrease of the Strouhal number for the smooth cylinder at $h^*/d^* = 0.05$, however, in their numerical experiment, the Strouhal number did not vanish ($St \approx 0.080$). In the present simulation, the correspondent computed value was of $St \approx 0.107$ (Table 2). An important conclusion is that for the smooth cylinder placed too close of the moving wall, i.e., at $h^*/d^* = 0.05$, the full interruption of von Kármán-type vortex shedding is not captured only using the active control technique by moving wall (Figure 7e–h).

The Strouhal number behavior was reported by Buresti and Lanciotti [12] for smooth and rough cylinders near a fixed ground at Reynolds numbers in the range from $Re = 8.5 \times 10^4$ to 3.0×10^5. The boundary layer thickness on the ground was about $\delta^*/d^* = 0.1$ at cylinder location. For the flow around the smooth cylinder, within the subcritical and critical regimes ($Re < 1.9 \times 10^5$), the critical gap-to-diameter ratio, h^*/d^*, was identified at 0.40, and the Strouhal number was estimated around of 0.20 for all ratios h^*/d^* greater than 0.40. The same results were obtained within the subcritical and critical regimes (in this situation, $Re < 1.4 \times 10^5$ because of the surface roughness effect) also for the rough cylinder. Although there is a lack of experimental data of Strouhal number for the flow around the smooth and rough cylinders near a moving wall, our numerical results agree basically with an experimental result of $St \approx 0.20$ for the cylinder near a fixed ground at large-gap regime. For the small-gap regime at $h^*/d^* < 0.20$, the Strouhal number decreases, and the boundary layer separation is delayed because of the combined effects of surface roughness and wall confinement (Table 2).

Figure 10a–h can be accompanied in sequence to better understand both interferences of surface roughness and wall confinement, which combined, completely destroy the orderly von Kármán vortex street. Instead, the Venturi effect interferes redrawing the negative shear layer parallel to ground plane behind the rough cylinder at $h^*/d^* = 0.05$. Figures 8d and 10e–h sketch the near wake pattern of the rough cylinder at $h^*/d^* = 0.05$, being the instants defined by points A–D in Figure 9b.

The difference of drag reduction between the study cases at $h^*/d^* = 0.05$ is explained because the base pressure increases for the rough cylinder (Figure 11b). It can be identified a lesser increase in the base pressure of the smooth cylinder (Figure 11a), which explains the difference about 36.3% in the drag force between them (Table 1).

Further numerical investigation will be carried out elsewhere to fully understand the relationship of the instantaneous surface pressure behavior with the mutual interaction between the two layers of opposite signals, when the complete interruption of vortex shedding is anticipated, for the rough cylinder at $h^*/d^* = 0.05$. Figures 8d and 10e–h give us some hints. Some small vortical structures observed in Figure 8d has been formed because of the two shear layers of opposite signals injecting vorticity at the rear part of the cylinder. As captured in our animations, the advection of vorticity of signals positive and negative creates those small vortical structures.

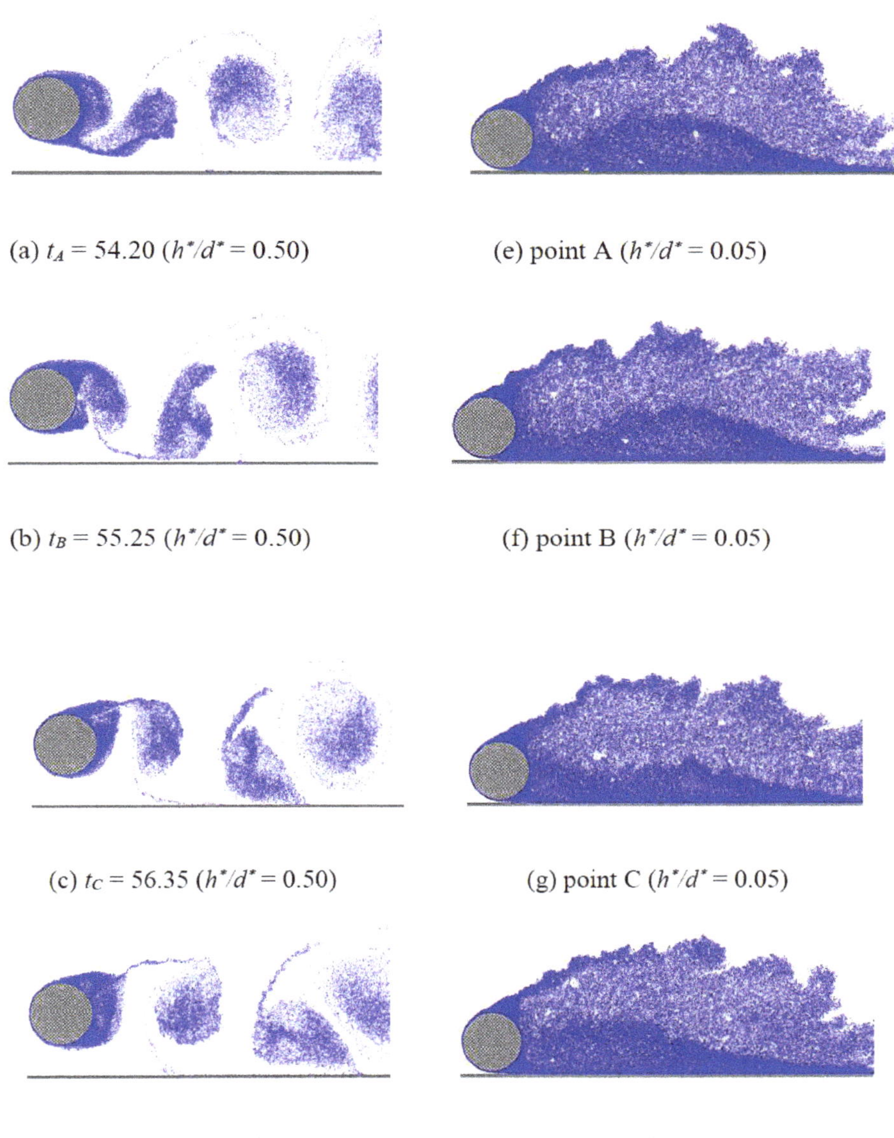

(a) $t_A = 54.20$ $(h^*/d^* = 0.50)$

(e) point A $(h^*/d^* = 0.05)$

(b) $t_B = 55.25$ $(h^*/d^* = 0.50)$

(f) point B $(h^*/d^* = 0.05)$

(c) $t_C = 56.35$ $(h^*/d^* = 0.50)$

(g) point C $(h^*/d^* = 0.05)$

(d) $t_D = 57.65$ $(h^*/d^* = 0.50)$

(h) point D $(h^*/d^* = 0.05)$

Figure 10. Near wake patterns for the slightly rough circular cylinder in the vicinity of a moving wall ($\varepsilon^*/d^* = 0.001$; $Re = 1.0 \times 10^5$).

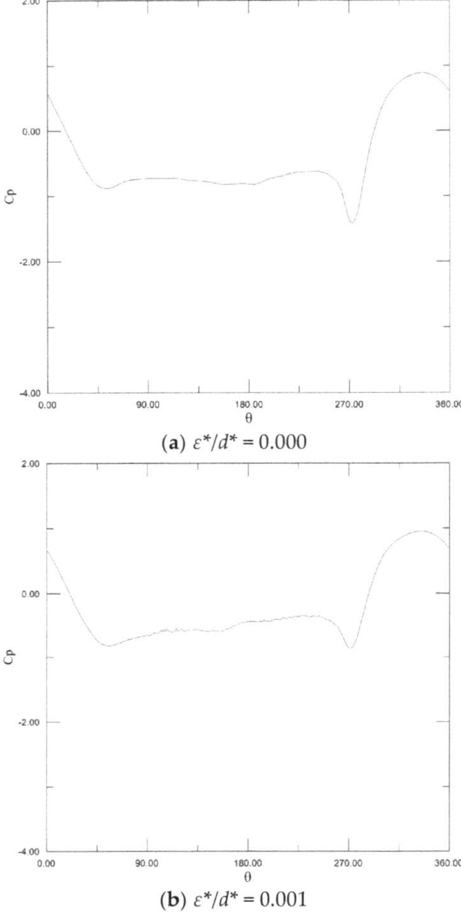

(a) $\varepsilon^*/d^* = 0.000$

(b) $\varepsilon^*/d^* = 0.001$

Figure 11. Time-averaged pressure distributions for the circular cylinder in the vicinity of a moving wall ($h^*/d^* = 0.05$; $Re = 1.0 \times 10^5$).

4. Conclusions

The present numerical study was addressed for the control and suppression of vortex shedding from a slightly rough bluff body in the vicinity of moving wall. A discrete vortex method implemented with a hybrid control technique of vortex shedding was employed. The association of active and passive control techniques of vortex shedding to study the two-dimensional flow past the rough circular cylinder placed too close to a moving wall (i.e., for the gap spacing of $h^*/d^* = 0.05$ in Figure 1) successfully captured the complete interruption of von Kármán-type vortex shedding (Figure 8d). The philosophy of our research line is to attain supercritical Reynolds number flow patterns starting from subcritical flows [11]. Achenbach [9] presented a classical $\overline{C}_D \times Re$ diagram (being $\overline{C}_D$ the mean drag coefficient) for the circular cylinder, and divided it into four flow regimes, i.e., subcritical ($Re < 2.0 \times 10^5 - 5.0 \times 10^5$), critical ($Re \cong 2.0 \times 10^5 - 5.0 \times 10^5$), supercritical ($2.0 \times 10^5 - 5.0 \times 10^5 < Re \leq 3.5 \times 10^6$) and transcritical ($Re > 3.5 \times 10^6$), being the latter nowadays called post-critical. The numerical results reported in the present work have been substantiated by experimental data of the flow past the smooth cylinder in the vicinity of a moving wall at upper-subcritical Reynolds number flow of $Re = 1.0 \times 10^5$ [25]. The key findings of the present study are summarized below.

(a) The most important result obtained for the rough cylinder is the decrease in its drag force and, as consequence, the delay in the boundary layer separation as the ratio h^*/d^* reduces (Table 2). The drag reduction is substantiated by an increase in the base pressure of the cylinder as compared as the smooth cylinder configuration (Figure 11a,b).

(b) The appearance of lift force pointing away from the moving wall is identified as the ratio h^*/d^* reduces (Table 1). This result is because of viscosity effect, which moves the cylinder front stagnation point downstream and creates an additional positive circulation contributing to increase the lift force (e.g., Figure 11a,b).

(c) The surface roughness effect on the cylinder at $h^*/d^* = 0.05$ is the key factor to anticipate the complete interruption of vortex shedding suppression, and, thus, the Strouhal number completely vanishes ($St = 0.0$ in Table 2); also two nearly parallel shear layers of opposite signals are created at the rear part of the cylinder (Figure 10e–h), being that the Venturi effect definitively contributes to redraw the negative shear layer parallel to the wall plane.

(d) The critical Strouhal number behavior for the rough cylinder at $h^*/d^* = 0.05$ is directly connected to a global change in its viscous wake structure (Figure 8d). In past work by Bimbato et al. [22], the full interruption of the formation of Von Kármán large-scale vortices was not captured for the flow past the smooth cylinder in the vicinity of a moving wall at $h^*/d^* = 0.05$ (Figure 8b,d can be used to comparison between test cases). Alcântara Pereira et al. [10] reported that the complete interruption of vortex shedding for the cylinder can be anticipated using a relative roughness size of $\varepsilon^*/d^* = 0.007$ at $h^*/d^* = 0.10$. Therefore, the present work also completes the past study of Alcântara Pereira et al. [10], now including discussions for the slightly rough cylinder with relative roughness size of $\varepsilon^*/d^* = 0.001$.

(e) The saturation state of a typical numerical simulation at dimensionless time of $t = 75$ is demonstrated through of the difference about 2% between Strouhal number values, one of them obtained from the inverse of the period $t_E - t_A$ (Figure 6a,b and Figure 9a,b), and other one computed from a Fast Fourier Transformation of the lift curve between $37.5 \leq t \leq 75$ (Figure 5a,d).

(f) The work also highlights vortex shedding partial suppression with representative reduction of both the Strouhal number (Table 1) and fluctuating drag and lift forces (Figure 5a,b).

(g) The present results suggest the applicability of the discrete vortex method with two-dimensional roughness model to study flows involving vortex shedding, and its control and suppression, from a bluff body in many practical problems, such as vibration of pipelines, transmission lines, suspension bridges, buildings, (semi-) submerged oil platform columns, heat exchangers tubes, periscopes and so on.

(h) The physical interpretations that have been presented in the present approach also contribute to report the sensitivity of the roughness model. It can be pointed that it is successfully captured and numerical simulations with acceptable accuracy for practical engineering applications can be performed.

Author Contributions: Conceptualization, M.A.d.O., C.L.d.A., and L.A.A.P.; data curation, M.A.d.O., P.G.d.M., and C.L.d.A.; formal analysis, C.L.d.A., A.M.B., and L.A.A.P.; funding acquisition, L.A.A.P.; investigation, M.A.d.O., P.G.d.M., C.L.d.A., and A.M.B.; methodology, C.L.d.A., A.M.B., and L.A.P.P.; project administration, A.M.B. and L.A.A.P.; resources, C.L.d.A. and L.A.A.P.; software using FORTRAN, C.L.d.A. and A.M.B.; validation, M.A.d.O., P.G.d.M., and C.L.d.A.; visualization, M.A.d.O., P.G.d.M., and C.L.d.A.; supervision, L.A.A.P.; writing—original draft, L.A.A.P.; and writing—review and editing, A.M.B. and L.A.A.P. All authors have read and agreed to the published version of the manuscript.

Funding: This research was supported by the FAPEMIG through the research project APQ-02175-14.

Conflicts of Interest: The authors declare no conflict of interest.

References

1. Bearman, P.W. Vortex shedding from oscillating bluff body. *Ann. Rev. Fluid Mech.* **1984**, *16*, 195–222. [CrossRef]
2. Gao, Y.; Fu, S.; Wang, J.; Song, L.; Chen, Y. Experimental study of the effects of surface roughness on the vortex-induce vibration of a flexible cylinder. *Ocean Eng.* **2015**, *103*, 40–54. [CrossRef]
3. Hirata, M.H.; Alcântara Pereira, L.A.; Recicar, J.N.; Moura, W.H. High Reynolds number oscillations of a circular cylinder. *J. Braz. Soc. Mech. Sci. Eng.* **2008**, *30*, 304–312. [CrossRef]
4. Hong, K.S.; Shah, U.M. Vortex-induced vibrations and control to marine risers: A review. *Ocean Eng.* **2018**, *152*, 300–315. [CrossRef]
5. Ma, W.; Liu, Q.; Macdonald, J.H.G.; Yan, X.; Zeng, Y. The effect of surface roughness on aerodynamic forces and vibrations for a circular cylinder in the critical Reynolds number range. *J. Wind Eng. Ind. Aerodyn.* **2019**, *187*, 61–72. [CrossRef]
6. Sun, H.; Ma, C.; Kim, E.O.; Nowakowski, G.; Bernitsas, M.M. Flow-induced vibration of tandem circular cylinders with selective roughness: Effect of spacing, damping and stiffness. *Eur. J. Mech. B/Fluids* **2019**, *74*, 219–241. [CrossRef]
7. Faure, T.M.; Dumas, L.; Montagnier, O. Numerical study of two-airfoil arrangements by a discrete vortex method. *Theor. Comput. Fluid Dyn.* **2020**, *34*, 79–103. [CrossRef]
8. Los Reis, J.H.; Alcântara Pereira, L.A. Particle-particle interactions in parallel computations. In Proceedings of the 16th Brazilian Congress of Thermal Sciences and Engineering, Vitoria, ES, Brazil, 7–10 November 2016.
9. Achenbach, E. Influence of surface roughness on the cross-flow around a circular cylinder. *J. Fluid Mech.* **1971**, *46*, 321–335. [CrossRef]
10. Alcântara Pereira, L.A.; Oliveira, M.A.; Moraes, P.G.; Bimbato, A.M. Numerical experiments of the flow around a bluff body with and without roughness model near a moving wall. *J. Braz. Soc. Mech. Sci. Eng.* **2020**, *42*, 129. [CrossRef]
11. Bimbato, A.M.; Alcântara Pereira, L.A.; Hirata, M.H. Development of a new Lagrangian vortex method for evaluating effects of surface roughness. *Eur. J. Mech. B/Fluids* **2019**, *74*, 291–301. [CrossRef]
12. Buresti, G.; Lanciotti, A. Vortex shedding from smooth and roughened cylinders in cross-flow near a plane surface. *Aeronaut. Q.* **1979**, *30*, 305–321. [CrossRef]
13. Buresti, G. The effect of surface roughness on the flow regime around circular cylinders. *J. Wind Eng. Ind. Aerodyn.* **1981**, *8*, 105–114. [CrossRef]
14. Fage, A.; Warsap, J.H. *The Effects of Turbulence and Surface Roughness on the Drag of a Circular Cylinder*; H.M. Stationary Office: London, UK, 1929.
15. Gupta, A.; Saha, A.K. Suppression of *vortex* shedding in flow around a square cylinder using control cylinder. *Eur. J. Mech. B/Fluids* **2019**, *76*, 276–291. [CrossRef]
16. Guven, O.; Farell, C.; Patel, V.C. Surface roughness effects on the mean flow past circular cylinders. *J. Fluid Mech.* **1980**, *98*, 673–701. [CrossRef]
17. Kareem, A.; Cheng, C.M. Pressure and force fluctuations on isolated roughened circular cylinders of finite height in boundary layer flows. *J. Fluids Struct.* **1999**, *13*, 907–934. [CrossRef]
18. van Hinsberg, N.P. The Reynolds number dependency of the steady and unsteady loading on a slightly rough circular cylinder: From subcritical up to high transcritical flow state. *J. Fluids Struct.* **2015**, *55*, 526–539. [CrossRef]
19. Yamagishi, Y.; Oki, M. Effect of groove shape on flow characteristics around a circular cylinder with grooves. *J. Vis.* **2004**, *7*, 209–216. [CrossRef]
20. Rashidi, S.; Hayatdavoodi, M.; Esfahani, J.A. Vortex shedding suppression and wake control. *Ocean Eng.* **2016**, *126*, 57–80. [CrossRef]
21. Bimbato, A.M.; Alcântara Pereira, L.A.; Hirata, M.H. Study of the vortex shedding flow around a body near a moving ground. *J. Wind Eng. Ind. Aerodyn.* **2011**, *99*, 7–17. [CrossRef]
22. Bimbato, A.M.; Alcântara Pereira, L.A.; Hirata, M.H. Suppression of vortex shedding on a bluff body. *J. Wind Eng. Ind. Aerodyn.* **2013**, *121*, 16–28. [CrossRef]
23. Lei, C.; Cheng, L.; Kavanagh, K. Re-examination of the effect of a plane boundary on force and vortex shedding of a circular cylinder. *J. Wind Eng. Ind. Aerodyn.* **1999**, *80*, 263–286. [CrossRef]

24. Li, Z.; Lan, C.; Jia, L.; Ma, Y. Ground effects on separated laminar flows past an inclined flat plate. *Theor. Comput. Fluid Dyn.* **2017**, *31*, 127–136. [CrossRef]
25. Nishino, T. Dynamics and Stability of Flow Past a Circular Cylinder in Ground Effect. Ph.D. Thesis, Faculty of Engineering Science and Mathematics, University of Southampton, Southampton, UK, 2007.
26. Roshko, A.; Steinolfson, A.; Chattoorgoon, V. Flow forces on a cylinder near a wall or near another cylinder. In Proceedings of the 2nd US Nation Conference on Wind Engineering Research, Fort Collins, CO, USA, 22–25 June 1975.
27. Zdravkovich, M.M. Forces on a circular cylinder near a plane wall. *Appl. Ocean Res.* **1985**, *7*, 197–201. [CrossRef]
28. Zdravkovich, M.M. *Flow around Circular Cylinders. Volume 2: Applications*; Oxford University Press: Oxford, UK, 2003.
29. Cui, E.; Zhang, X. Ground effect aerodynamics (Chapter 18). In *Encyclopedia of Aerospace Engineering*; John Wiley & Sons, Ltd.: Hoboken, NJ, USA, 2010.
30. Zhang, X.; Toet, W.; Zerihan, J. Ground effect aerodynamics of race cars. *ASME Appl. Mech. Rev.* **2006**, *59*, 33–49. [CrossRef]
31. Blevins, R.D. *Applied Fluid Dynamics Handbook*; Van Nostrand Reinhold Co.: Hoboken, NJ, USA, 1984.
32. Katz, J.; Plotkin, A. *Low Speed Aerodynamics: From Wing Theory to Panel Methods*; McGraw Hill Inc.: New York, NY, USA, 1991.
33. Kundu, P.K. *Fluid Mechanics*; Academic Press: Cambridge, UK, 1990.
34. Leonard, A. Vortex methods for flow simulation. *J. Comp. Phys.* **1980**, *37*, 289–335. [CrossRef]
35. Chorin, A.J. Numerical study of slightly viscous flow. *J. Fluid Mech.* **1973**, *57*, 785–796. [CrossRef]
36. Alcântara Pereira, L.A.; Hirata, M.H.; Silveira Neto, A. Vortex method with turbulence sub-grid scale modeling. *J. Braz. Soc. Mech. Sci. Eng.* **2003**, *25*, 140–146.
37. Helmholtz, H. On integrals of the hydrodynamics equations which express vortex motion. *Philos. Mag.* **1867**, *33*, 485–512. [CrossRef]
38. Lesieur, M.; Métais, O. New trends in large eddy simulation of turbulence. *Annu. Rev. Fluid Mech.* **1996**, *28*, 45–82. [CrossRef]
39. Alcântara Pereira, L.A.; Hirata, M.H.; Manzanares Filho, N. Wake and aerodynamics loads in multiple bodies: Application to turbomachinery blade rows. *J. Wind Eng. Ind. Aerodyn.* **2004**, *92*, 477–491. [CrossRef]
40. Shintani, M.; Akamatsu, T. Investigation of two-dimensional discrete vortex method with viscous diffusion model. *CFD J.* **1994**, *3*, 237–254.
41. Gerrard, J.H. The mechanics of the formation region of vortices behind bluff bodies. *J. Fluid Mech.* **1966**, *25*, 401–413. [CrossRef]

Article

Analysis of Aeroacoustic Properties of the Local Radial Point Interpolation Cumulant Lattice Boltzmann Method

Mohsen Gorakifard [1], Clara Salueña [1,*], Ildefonso Cuesta [1] and Ehsan Kian Far [2]

1   Department of Mechanical Engineering, ETSEQ, Rovira i Virgili University, Països Catalans, 26, 43007 Tarragona, Spain; mohsen.goraki@urv.cat (M.G.); ildefonso.cuesta@urv.cat (I.C.)

2   Department of Mechanical Engineering, The University of Manchester, Oxford Rd, Manchester M13 9PL, UK; ehsan.kianfar@manchester.ac.uk

*   Correspondence: clara.saluena@urv.cat

Abstract: The lattice Boltzmann method (LBM) has recently been used to simulate wave propagation, one of the challenging aspects of wind turbine modeling and simulation. However, standard LB methods suffer from the instability that occurs at low viscosities and from its characteristic lattice uniformity, which results in issues of accuracy and computational efficiency following mesh refinement. The local radial point interpolation cumulant lattice Boltzmann method (LRPIC-LBM) is proposed in this paper to overcome these shortcomings. The LB equation is divided into collision and streaming steps. The collision step is modeled by the cumulant method, one of the stable LB methods at low viscosities. In addition, the streaming step, which is naturally a pure advection equation, is discretized in time and space using the Lax–Wendroff scheme and the local radial point interpolation method (RPIM), a mesh free method. We describe the propagation of planar acoustic waves, including the temporal decay of a standing plane wave and the spatial decay of a planar acoustic pulse. The analysis of these specific benchmark problems has yielded qualitative and quantitative data on acoustic dispersion and dissipation, and their deviation from analytical results demonstrates the accuracy of the method. We found that the LRPIC-LBM replicates the analytical results for different viscosities, and the errors of the fundamental acoustic properties are negligible, even for quite low resolutions. Thus, this method may constitute a useful platform for effectively predicting complex engineering problems such as wind turbine simulations, without parameter dependencies such as the number of points per wavelength N_{ppw} and resolution σ or the detrimental effect caused by the use of coarse grids found in other accurate and stable LB models.

Keywords: local radial point interpolation cumulant LBM; aeroacoustics; dispersion; dissipation; wind turbine

Citation: Gorakifard, M.; Salueña, C.; Cuesta, I.; Far, E.K. Analysis of Aeroacoustic Properties of the Local Radial Point Interpolation Cumulant Lattice Boltzmann Method. *Energies* **2021**, *14*, 1443. https://doi.org/10.3390/en14051443

Academic Editor: Robert Castilla

Received: 11 January 2021
Accepted: 2 March 2021
Published: 6 March 2021

Publisher's Note: MDPI stays neutral with regard to jurisdictional claims in published maps and institutional affiliations.

1. Introduction

Evidence of early sailing boats on the Nile and of Persian pumps and mills from the first century B.C. shows humans have been interested in Wind Energy since ancient times [1]. In general, a wind turbine is defined as a device which converts the wind's kinetic energy into electrical energy [2,3]. It plays a key role on producing intermittent renewable energy and implementing a strategy to lower costs and reducing the reliance on fossil fuels [4,5]. Wind turbines have unique aerodynamic and aeroacoustic behavior that makes their prediction most challenging [6,7], particularly their simulation needs an enormous number of grid points or cells, and long enough time samples [8]. Researchers and centers such as the National Renewable Energy Laboratory (NREL) and the National Wind Technology Center (NWTC) have initiated multi-year programs on aeroacoustic wind turbine modeling [9] to develop efficient and appropriate computational aeroacoustic (CAA) implementations. Among particular issues specified to wind turbine problems, the propagation of sound is always a significant computational challenge [10,11]. With this aim, different numerical approaches were developed in the field of computational aeroacoustics. Tam [12] and Wells et al. [13] proposed popular numerical schemes such as compact and non-compact optimized schemes like the high-order compact difference scheme [14] and the dispersion-relation-preserving (DRP) scheme [15]. Cheong et al. proposed grid-optimized dispersion-relation-preserving (GODRP) schemes for aeroacoustic simulations [16]. Due to the huge cost of CAA simulations, hybrid methods using two sets of equations, one for the flow and another one for the acoustic disturbance field were developed [17].

Direct aerostatic simulations are computer-intensive due to the small ratio between sound pressure and pressure variation as a whole, the spreading of acoustic fields over a large area, and the time-consuming nature of traditional methods [18]. As an illustration, the direct numerical simulation of waves using Navier–Stokes equations requires schemes of fifth-order accuracy in space and fourth order accuracy in time [19,20]. Therefore, the lattice Boltzmann method (LBM), an explicit time marching scheme [21], has been widely used as an alternative to simulate sound wave propagation due to its kinetic nature, relative simplicity of implementation and parallelization. The LBM as a mesoscopic method uses probability density functions (probability of finding particles within a certain range of velocities at a certain range of locations at a given time) to model the momentum distribution in discrete space, thereby economizing computer resources [22]. Buick et al. [23] and Dellar et al. [24] studied sound wave propagation using LBM and achieved acceptable results. Bres et al. [25] and Gorakifard et al. [26] presented the dissipation and dispersion of acoustic waves using the BGK-LBM and the cumulant LBM, respectively. Furthermore, a regularized method for the BGK-LBM [27] and the recursive and regularized LBM (LBM-rrBGK) [28,29] have been developed to model wave propagation.

One of the drawbacks of the LBM is lattice uniformity, originated from symmetric lattice velocity models such as the square and cubic lattice meshes in 2D/3D simulations [30]. Lattice uniformity causes the streaming step to occur at uniform neighboring grid points. Thus, the LBM is only applied to uniform meshes, whereas issues of accuracy and computational efficiency mainly affect simulations of problems that require non-uniform meshes. For example, numerical simulations with curved and irregular boundaries, common in wind turbine simulations, encounter difficulties when fitting the grids to the boundaries or adapting complex computational domains. Grid refinement schemes or adaptive LBM can help to simulate curved and moving boundaries more accurately [31]. Wood [32] used refinement in LBM simulations to analyze wind energy and utilized adaptive LBM for moving boundaries [33]. However, these schemes are associated with higher computational costs and even additional perturbations in acoustical problems [34].

Generally, the methods commonly used along with the LBM on non-uniform meshes include three distinct categories [35]. The first is the interpolation-supplemented LBM (IS-LBM) [36,37]. This method adds an interpolation step to the collision and streaming steps of the LBM. Two major drawbacks of the IS-LBM are its inefficiency due to the need to interpolate at each time-step, and the appearance of negative particle distributions [38]. The second method is the combination of the LBM with the finite difference method (FDLBE) [39], finite volume methods (FVLBE) [40–42], or finite element methods (FELBE) [43–45] used to stabilize the computation. The third method is the Taylor-series expansion and least-squares-based lattice Boltzmann method (TLLBM) [46,47] instead of direct interpolation. These methods, the implementation of which is somewhat simple, use continuous distribution functions in physical space.

Although the above numerical schemes show robustness for complex problems, they are affected by the inherent shortcomings of using meshes in numerical methods, such as the enormous cost of generating meshes, the low level of stress accuracy in fluid–structure interaction simulations (FSI) [48], obstacles in the adaptive analysis, and limitations in simulations of physical phenomena with singular, or nearly singular behavior. Mesh-free or meshless methods have been devised to eliminate mesh-related problems [49]. The MFree method based on Liu's definition [50] is "a method that generates a system of algebraic equations for the entire problem domain without consideration of a predefined mesh". This means that the method needs a set of scattered nodes inside the problem domain and on the boundaries of the domain called field nodes. In addition, the relationship between the nodes for the interpolation or approximation of the unknown field variables is not required [49]. Some of these well-known methods are the local point interpolation method (LPIM) [51], the local radial point interpolation method (LRPIM) [52], and the meshless local Petrov–Galerkin method (MLPG) [53,54].

The coupling of LBM and MFree methods has achieved acceptable results in some applications [55–58]; however, this approach is at an early stage of development and must be improved to address instability in low viscosities and high Re numbers. A key point in the stability and the accuracy of these methods is the collision operator, which is not remarkable in early LB methods such as the BGK model and the multi-relaxation times (MRT) model [59], both of which also violate the principle of Galilean invariance. Therefore, using a suitable collision operator to simulate complex engineering problems accurately has been recognized as a necessity. The advent of the more stable cumulant LBM [60,61] could dramatically improve MFree-LB methods and contribute to their advancement as powerful numerical tools for complex simulations.

The aim of this paper is to study the capability of the local radial point interpolation cumulant LBM (LRPIC-LBM) to simulate the propagation of planar acoustic waves, including the temporal decay of a standing plane wave and the spatial decay of a planar acoustic pulse of Gaussian shape and calculate the deviation from theoretical results, and to determine whether this method might be useful for wind turbine problems. The LB equation is deconstructed into collision and streaming parts. The collision step is performed by means of the cumulant method. The streaming step, which represents a pure advection equation, is discretized first in time using the Lax–Wendroff scheme, and then in space using the local radial point interpolation method (RPIM). Thus, this paper is arranged in the following sections; Section 2 is devoted to a brief summary of our new LB method. Section 2.1 presents the collision part referred to as cumulant collision step. Section 2.2 is devoted to the second most important part of the LBM, the streaming step. The Mfree shape function construction (RPIM shape function), time discretization (Lax–Wendroff scheme), and space discretization (LRPIM) are discussed in Sections 2.2.1–2.2.3. Section 3 reports the results of the LRPIC-LBM. Particularly, the planar standing wave and planar pulse wave propagation results are discussed in Sections 3.1 and 3.2.

2. The Lattice Boltzmann Method

The lattice Boltzmann method (LBM) is obtained from the kinetic theory of gases. In the LBM, a key point for modeling the momentum distribution in discrete space is the use of probability density functions [22]. The lattice Boltzmann equation without an external force is

$$f_i\left(x + e_{xi}c\Delta t, y + e_{yi}c\Delta t, t + \Delta t\right) - f_i\left(x, y, t\right) = \Omega_i \tag{1}$$

where f_i, Ω_i, and c are the particle distribution function, the collision operator, and the lattice speed, respectively.

In general, the LBM consists of collision and streaming steps. In the local radial point interpolation cumulant LBM (LRPIC-LBM), the cumulant method is used for the collision part and the local radial point interpolation method (RPIM) is used for the streaming parts.

2.1. Collision Step

The cascade method has been proposed [62,63] as a way to overcome the instability problems and modeling artifacts of previous LB methods. However, it is hindered by the effects of lower order moments over higher order moments. The cumulant method, presented by Seeger to solve the Boltzmann Equation [64,65], effectively resolves these issues. Cumulants can be efficiently generated from central moments. The cumulant method for solving the LBM is briefly described in this section.

The probability density function (PDF) [22] is

$$f(\xi,\eta) = \sum_{ij} f(\xi_i,\eta_j)\delta(\xi-\xi_i)\delta(\eta-\eta_i),\tag{2}$$

where f is a probability mass function (PMF) and ξ,η are discrete random variables with ranges $R_\xi = \{\xi_1,\xi_2,\ldots\}$, $R_\eta = \{\eta_1,\eta_2,\ldots\}$ based on microscopic velocities in x,y directions.

Using the moment-generating function, $M(\Xi,H) = \sum_{ij} f(\xi_i,\eta_j)e^{\Xi\xi_i}e^{H\eta_j}$, the moments can be determined without any discontinuity issues as [66]

$$\mu'_{\xi^m\eta^n} = \left.\frac{\partial^m\partial^n}{\partial\Xi^m\partial H^n}M(\Xi,H)\right|_{\Xi=H=0},\tag{3}$$

where H,Ξ are the normalized wave numbers.

It is a good idea to shift the moment-generating function into the moving reference frame of the fluid to reduce the Galilean invariant issues in the collision process. Therefore, using the central moment-generating function, $\widehat{M}(\Xi,H) = e^{-\Xi u/c - Hv/c}M(\Xi,H)$, the central moments can be defined as [67]

$$\mu_{\xi^m\eta^n} = \left.\frac{\partial^m\partial^n}{\partial\Xi^m\partial H^n}\widehat{M}(\Xi,H)\right|_{\Xi=H=0}.\tag{4}$$

The cumulants are calculated using the logarithm form of the moment-generating function [68] as follows:

$$c_{\xi^m\eta^n} = \left.\frac{\partial^m\partial^n}{\partial\Xi^m\partial H^n}\ln(M(\Xi,H))\right|_{\Xi=H=0}.\tag{5}$$

Each cumulant relaxes with an individual relaxation rate [69].

$$c^*_{\xi^m\eta^n} = c_{\xi^m\eta^n} + \omega_{\xi^m\eta^n}(c^{eq}_{\xi^m\eta^n} - c_{\xi^m\eta^n}),\tag{6}$$

where $c^{eq}_{\xi^m\eta^n}$ are the cumulants of the equilibrium state.

The sound speed and the kinematic viscosity are defined as $c_s = \Delta x/(\sqrt{3}\Delta t)$ and $v = \Delta t c_s^2(1/\omega - 0.5)$, respectively.

2.2. Streaming Step

To model the streaming step, the second most important part of the LBM, a pure advection equation is normally solved from a Lagrangian approach within uniform structured meshes with CFL numbers equal to one. However, considering the Eulerian perspective for the calculation can effectively resolve this step when meshes are non-uniform and unstructured. The pure advection equation is

$$\frac{\partial f_i}{\partial t} + c_{i,\alpha}\frac{\partial f_i}{\partial x_\alpha} = 0,\tag{7}$$

One alternative is a semi-discrete formulation with time and space derivatives discretized separately. Thus, Equation (7) is discretized using the explicit Lax–Wendroff scheme in time, followed by the local radial point interpolation method (LRPIM) in space. In addition, it is important to approximate the unknown field functions using trial (shape) functions as an approximate solution for the partial differential equation.

2.2.1. MFree Shape Function Construction—Radial Point Interpolation Shape Functions

Radial point interpolation method (RPIM) shape functions were developed to circumvent the singularity problem arising in the point interpolation method (PIM). The RPIM interpolation augmented with polynomials is

$$f^h(x,t) = \sum_{i=1}^{n} R_i(x)a_i(t) + \sum_{j=1}^{m} p_j(x)b_j(t) = R^T(x)a(t) + p^T(x)b(t) \tag{8}$$

where $R_i(x)$ is a radial basis function (RBF), and $p_j(x)$ is monomial in the coordinate space $x^T = [x, y]$: $R^T = [R_1(x)\, R_2(x)\, \cdots\, R_n(x)]$ and $p^T = [p_1(x)\, p_2(x)\, \cdots\, p_m(x)]$. Parameters n and m are the number of RBFs and polynomial basis functions. Variables a_i and b_j are time dependent unknown coefficients. It should be noted that the independent variable in RBF $R_i(x)$ is the distance between the point of interest x and a node at x_i.

Different radial basis functions (RBF), and their characteristics have been studied extensively in the meshless RPIM. In this paper, the multi-quadrics (MQ) function is used as

$$R_i = \left(r_i^2 + (\alpha_c d_c)^2\right)^q \tag{9}$$

where $\alpha_c = 1.0$, $q = 1.03$, and $d_c = 3.0$.

To determine the $n + m$ unknown coefficients in Equation (8), some specific constraint equations and the Kronecker delta function property are dictated. These constraints are

$$\sum_{i=1}^{n} p_j(x_i)a_i(t) = P_m^T a(t) = 0, \qquad j = 1, 2, ..., m \tag{10}$$

where

$$P_m^T = \begin{bmatrix} 1 & 1 & \cdots & 1 \\ x_1 & x_2 & \cdots & x_n \\ y_1 & y_2 & \cdots & y_n \\ \vdots & \vdots & \ddots & \vdots \\ p_m(x_1) & p_m(x_2) & \cdots & p_m(x_n) \end{bmatrix} \tag{11}$$

Thus, the approximation function can be obtained as

$$f^h(x,t) = \sum_{i=1}^{n} \phi_i(x)f_i(t) = \Phi(x)F(t) \tag{12}$$

where F is a vector containing the nodal values of the distribution function and Φ is a vector containing the first n components of the $\tilde{\Phi}$ vector

$$\tilde{\Phi} = \left[R^T p^T\right]G^{-1} \tag{13}$$

where

$$G = \begin{bmatrix} R_0 P_m \\ P_m^T 0 \end{bmatrix}, \quad R_0 = \begin{bmatrix} R_1(r_1) & R_2(r_1) & \cdots & R_n(r_1) \\ R_1(r_2) & R_2(r_2) & \cdots & R_n(r_2) \\ \vdots & \vdots & \ddots & \vdots \\ R_1(r_n) & R_2(r_n) & \cdots & R_n(r_n) \end{bmatrix}, \quad r_k = \sqrt{(x_k - x_i)^2 + (y_k - y_i)^2}. \tag{14}$$

2.2.2. Semi-Discrete Formulation—Time Discretization

The Taylor series expansion of the particle distributions is

$$f_i^{n+1} = f_i^n + \Delta t \frac{\partial f_i}{\partial t}\bigg|^n + \frac{\Delta t^2}{2}\frac{\partial^2 f_i}{\partial t^2}\bigg|^n + O(\Delta t^3), \tag{15}$$

where here n refers to the time step. Substituting the time derivatives in terms of the t derivatives up to second order results in the time discretization of Equation (7) based on the Lax–Wendroff scheme,

$$f_i^{n+1} = f_i^n - \Delta t c_{i,\alpha} \frac{\partial f_i^n}{\partial x_\alpha} + \frac{\Delta t^2}{2} c_{i,\alpha} c_{i,\beta} \frac{\partial^2 f_i^n}{\partial x_\alpha \partial x_\beta}. \tag{16}$$

2.2.3. Semi-Discrete Formulation—Space Discretization

The local radial point interpolation method (LRPIM) was developed to avoid the side effects of using global background cells in the global weak-form. In this method, the numerical integration is performed within the local domain consisting of a set of distributed nodes. The LRPIM is based on the RPIM shape functions with the delta function property. The main advantage of the LRPIM is the excellent interpolation stability of RBFs.

MFree local weak-form methods use the weak form of the problem obtained from the method of weighted residuals (MWR). The weighted residual statement of Equation (16) on the local domain Ω_I of point I bounded by Γ_I is posed as

$$\int_{\Omega_I} w_I f_i^{n+1} \, d\Omega = \int_{\Omega_I} w_I f_i^n \, d\Omega - \Delta t \int_{\Omega_I} w_I c_{i,\alpha} \frac{\partial f_i^n}{\partial x_\alpha} \, d\Omega + \frac{\Delta t^2}{2} \int_{\Omega_I} w_I c_{i,\alpha} c_{i,\beta} \frac{\partial^2 f_i^n}{\partial x_\alpha \partial x_\beta} \, d\Omega \tag{17}$$

where w_I is the local weight function of node I considered as

$$w_I(r) = \begin{cases} \frac{2}{3} - 4r^2 + 4r^3 & r \le 0.5 \\ \frac{4}{3} - 4r + 4r^2 - \frac{4}{3}r^3 & 0.5 < r \le 1 \\ 0 & r > 1 \end{cases} \tag{18}$$

where $r = |x - x_i|/d^{\max}$ and $d^{\max}$ is the radius of the compact support. Equation (17) is applied to all nodes in the problem domain.

To work with a continuous approximate solution, it is necessary to decrease the differentiation requirements of the unknown in the weighted residual statement by employing integration by parts in Equation (17),

$$\int_{\Omega_I} w_I f_i^{n+1} \, d\Omega = \int_{\Omega_I} w_I f_i^n \, d\Omega - \int_{\Omega_I} \left(\Delta t w_I c_{i,\alpha} \frac{\partial f_i^n}{\partial x_\alpha} + \frac{\Delta t^2}{2} c_{i,\alpha} c_{i,\beta} \frac{\partial w_I}{\partial x_\beta} \frac{\partial f_i^n}{\partial x_\alpha} \right) d\Omega + \frac{\Delta t^2}{2} \int_{\Gamma_I} w_I c_{i,\alpha} c_{i,\beta} \frac{\partial f_i^n}{\partial x_\alpha} n_\beta \, d\Gamma \tag{19}$$

where Γ_I is the boundary of the local domain Ω_I and n_β is the unit outward normal vector.

Substitution of the approximate solution given in Equations (12) into the weak form given by Equation (19) leads to

$$\sum_{J=1}^{N_I} M_{IJ} f_{i,J}^{n+1} = \sum_{J=1}^{N_I} \left[M_{IJ} + K_{i,IJ} \right] f_{i,J}^n \tag{20}$$

where M_{IJ} and $K_{i,IJ}$ are the nodal mass and stiffness matrix, respectively, defined as

$$M_{IJ} = \int_{\Omega_I} w_I \Phi_J \, d\Omega \tag{21}$$

$$K_{i,IJ} = - \int_{\Omega_I} \left(\Delta t w_I + \frac{\Delta t^2}{2} c_{i,\beta} \frac{\partial w_I}{\partial x_\beta} \right) c_{i,\alpha} \frac{\partial \Phi_J}{\partial x_\alpha} \, d\Omega + \frac{\Delta t^2}{2} \int_{\Gamma_I} w_I c_{i,\alpha} \frac{\partial \Phi_J}{\partial x_\alpha} c_{i,\beta} n_\beta \, d\Gamma \tag{22}$$

Thus, the global equation system for all nodes in the entire domain is obtained as

$$M f_i^{n+1} = [M + K_i] f_i^n \tag{23}$$

where M, K, and f_i are the global mass matrix, stiffness matrix, and particle distribution vector, respectively. This system has N equations with N unknowns which is solved separately for each direction.

To numerically evaluate the area and the curve integrals in Equations (21) and (22) the Gauss quadrature scheme is used as follows

$$M_{IJ} = \sum_{k=1}^{n_g} \widetilde{w}_k w_I(x_k) \Phi_J(x_k) \left| J^{\Omega_I} \right| \tag{24}$$

$$K_{i,IJ} = -\sum_{k=1}^{n_g} \widetilde{w}_k \left(\Delta t w_I(x_k) + \frac{\Delta t^2}{2} c_{i,\beta} \frac{\partial w_I}{\partial x_\beta} \Big|_{x_k} \right) c_{i,\alpha} \frac{\partial \Phi_J}{\partial x_\alpha} \Big|_{x_k} \left| J^{\Omega_I} \right| + \frac{\Delta t^2}{2} \sum_{k=1}^{n_g^b} \widetilde{w}_k w_I(x_k) c_{i,\alpha} \frac{\partial \Phi_J}{\partial x_\alpha} \Big|_{x_k} c_{i,\beta} n_\beta \left| J^{\Gamma_I} \right| \tag{25}$$

where n_g and n_g^b are the total number of Gauss points in the quadrature domain and boundaries, $\widetilde{w}_k$ is the Gauss weight factor for Gauss point x_k, J^{Ω_I} and J^{Γ_I} are the Jacobian matrix for the domain and boundary integrations, respectively.

3. Results and Discussion

One of the complicated phenomena that has recently received major interest from researchers using the LBM is wind turbine aeroacoustics [32,33,70,71], which can be directly simulated without additional computational cost. In this section, our aim is to demonstrate the standard analysis procedure using the local radial point interpolation cumulant lattice Boltzmann method (LRPIC-LBM) to predict acoustic properties for benchmark cases. Thus, numerical studies are conducted for the propagation of planar acoustic waves, concentrating on numerical dissipation and dispersion. Considering the total pressure equation as $p = p_0 + p'$, where p_0 is the atmospheric pressure and p' is the acoustic pressure, a lossy wave equation is [72]

$$\left(1 + \tau_s \frac{\partial}{\partial t} \right) \nabla^2 p' = \frac{1}{c_s^2} \frac{\partial^2 p'}{\partial t^2} \tag{26}$$

where

$$\tau_s = \frac{1}{\rho_0 c_s^2} \left(\frac{4}{3} \eta + \eta_B \right).$$

where η is the coefficient of shear viscosity, and η_B is the coefficient of bulk viscosity. A quantitative assessment of the method's two setups, including the temporal decay of a standing plane wave in a periodic domain and the spatial decay of a propagating planar acoustic pulse of Gaussian shape for regular and irregular nodal distributions (shown in Figure 1a,b) are discussed below. It should be noted that the default nodal arrangement is a regular distribution; the base units are in the LB system.

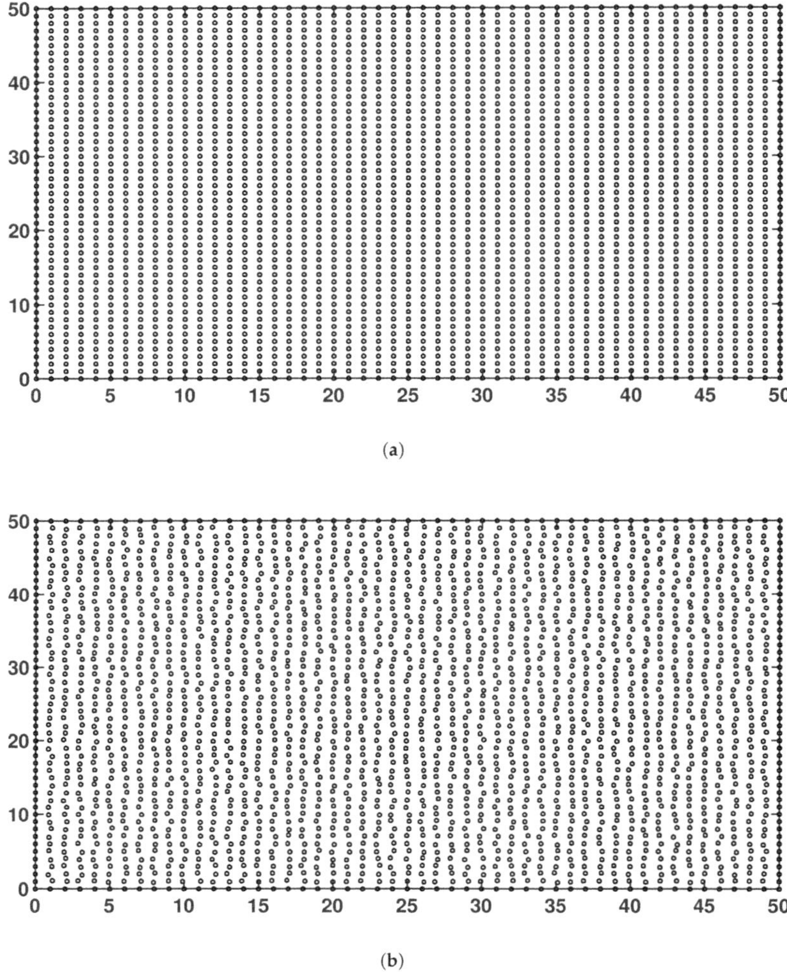

Figure 1. Nodal arrangement for the propagation of planar acoustic waves in LRPIC-LBM simulations: (**a**) Regular nodes. (**b**) Irregular nodes.

3.1. Planar Standing Wave

As an initial case study, we performed a temporal analysis on a standing plane acoustic wave in a periodic domain. The dissipation and dispersion relations based on the temporal analysis of Equation (26) are [25]

$$c_T = c_s\sqrt{1 - \left(\frac{k\nu}{c_s}\right)^2} \tag{27a}$$

$$\alpha_T = k^2\nu \tag{27b}$$

where k is the wave number. The assumptions for this set-up are presented in Table 1. They were chosen as in reference [26] for ease of comparison.

Table 1. Parameters for the planar standing wave

Variables	$p'(x,y,0)$	ρ'	u'	v'	A
Description	$A\sin\left(\frac{2\pi x}{\lambda}\right)$	$\frac{p'}{c_s^2}$	$\frac{p'}{\rho_0 c_s}$	0	$10^{-3}p_0$

The acoustic pressure at time t is [26]

$$p'(x,y,t) = A\exp[-\alpha_T t]\sin[k(x - c_T t)] \tag{28}$$

It should be noted that the results of the temporal analysis can be considered as a function of the number of points per wavelength $N_{ppw} = \lambda/\Delta x$ or the non-dimensional wave-number $k\Delta x = 2\pi/N_{ppw}$.

In accordance with the concepts discussed in [26] to study the accuracy of the propagation of waves using the local radial point interpolation cumulant lattice Boltzmann method (LRPIC-LBM), various viscosities and different resolutions (i.e., the number of points per wavelength) were studied. Figure 2 shows the acoustic pressure time history for $\nu = 1.0 \times 10^{-2}$ $(\frac{\Delta x^2}{\Delta t})$ with $N_{ppw} = 12$ points per wavelength. The analytical result is represented by a solid black line, whereas the cumulant LBM and the LRPIC-LBM are shown with a red dashed line and a blue five-pointed star, respectively. The deviations of the numerical phase speed and the temporal dissipation rate from the theoretical values are less than one percent for the BGK LBM [25] and the cumulant LBM [26] for resolutions lower than 12 points per wavelength. However, the LRPIC-LBM exhibits even better behavior in predicting the acoustic pressure of the analytical values.

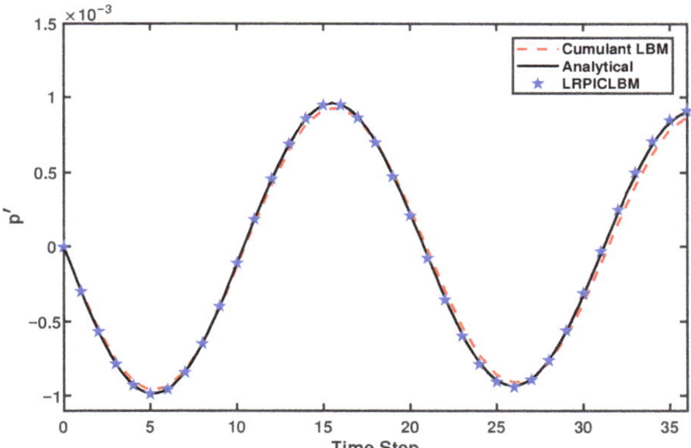

Figure 2. Acoustic pressure $(\frac{kg}{\Delta x \Delta t^2})$ versus time step (Δt) for $\nu = 1.0 \times 10^{-2}$ $(\frac{\Delta x^2}{\Delta t})$ with $N_{ppw} = 12$ points per wavelength.

The acoustic pressure time history for $\nu = 1.0 \times 10^{-4}$ $(\frac{\Delta x^2}{\Delta t})$ with $N_{ppw} = 12$ points per wavelength is presented in Figure 3. It shows the comparison between the analytical solution, the cumulant LBM and the LRPIC-LBM numerical solution at low viscosities. One of the most problematic issues in the BGK LBM is the low viscosity limit, which makes the solution unstable. However, the cumulant LBM, with phase speed and temporal dissipation rate errors of less than 1 percent, as in reference [26], did not present difficulties at low viscosity. Although both approaches were successful, the new method exhibited better performance, closely following the theoretical result for the same viscosity value.

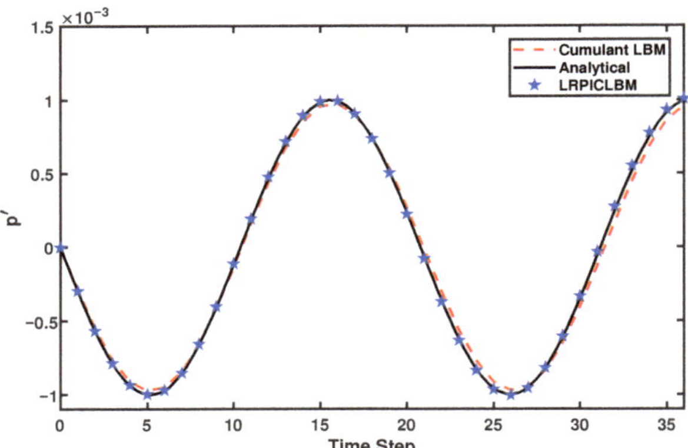

Figure 3. Acoustic pressure ($\frac{kg}{\Delta x \Delta t^2}$) versus time step ($\Delta t$) for $\nu = 1.0 \times 10^{-4}$ ($\frac{\Delta x^2}{\Delta t}$) with $N_{ppw} = 12$ points per wavelength.

An important characteristic highlighted by some researchers [25,26] is that these numerical deviations are only a function of N_{ppw} and are independent of other parameters such as frequency and viscosity. They found that the errors of the BGK [25] and the cumulant LBM [26] are about 7 percent for $N_{ppw} = 4$. However, the acoustic pressure time history for $\nu = 1.0 \times 10^{-2}$ ($\frac{\Delta x^2}{\Delta t}$) with $N_{ppw} = 4$ points per wavelength illustrated in Figure 4 for the analytical solution, the cumulant LBM and the LRPIC-LBM reveals that the deviation is less than 2 percent for the LRPIC-LBM, with $\Delta t = 0.25$. Thus, the LRPIC-LBM is much more successful in predicting theoretical results even with a low number of points per wavelength.

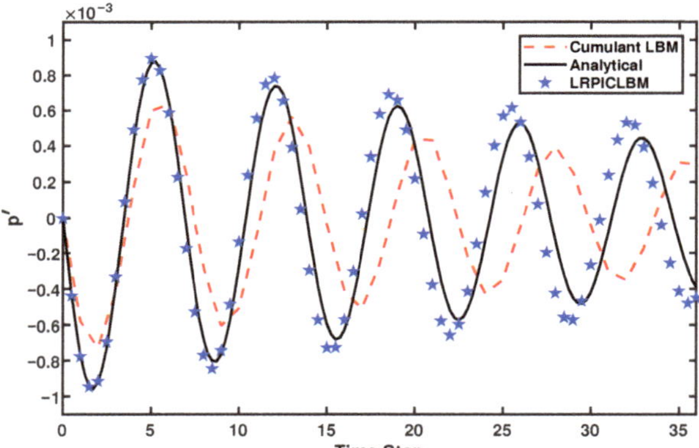

Figure 4. Acoustic pressure ($\frac{kg}{\Delta x \Delta t^2}$) versus time step ($\Delta t$) for $\nu = 1.0 \times 10^{-2}$ ($\frac{\Delta x^2}{\Delta t}$) with $N_{ppw} = 4$ points per wavelength with $\Delta t = 0.25$.

The choice of the time step size has an influence on the accuracy and stability of the solution. A smaller time step leads to more accurate results, especially in a hyperbolic partial differential Equation (PDE). To minimize the phase speed and dissipation rate errors, the time step was reduced. Figure 5 shows the acoustic pressure time history for $\nu = 1.0 \times 10^{-2}$ ($\frac{\Delta x^2}{\Delta t}$) with $N_{ppw} = 4$ points per wavelength and $\Delta t = 0.1$. The LRPI-CLBM replicates the analytical results with negligible errors. Therefore, this method makes it possible to predict wave motion more accurately with no dependency on the number of points per wavelength N_{ppw}.

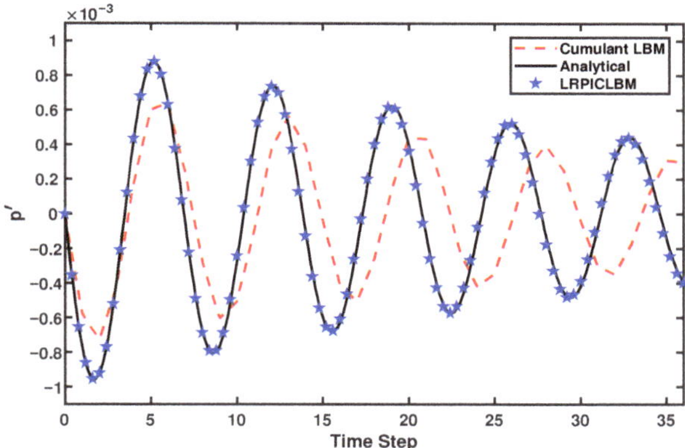

Figure 5. Acoustic pressure ($\frac{kg}{\Delta x \Delta t^2}$) versus time step ($\Delta t$) for $\nu = 1.0 \times 10^{-2}$ ($\frac{\Delta x^2}{\Delta t}$) with $N_{ppw} = 4$ points per wavelength with $\Delta t = 0.1$.

Although the results of the propagation of acoustic waves for regular nodal distributions were good, it is important to adequately estimate accuracy when considering irregular nodes (Figure 1b). The acoustic pressure time history for $\nu = 1.0 \times 10^{-2}$ ($\frac{\Delta x^2}{\Delta t}$) with $N_{ppw} = 12$ points per wavelength is presented in Figure 6. The results show that the local radial point interpolation cumulant lattice Boltzmann method (LRPIC-LBM) with irregular nodal distributions again very closely reproduced the analytical acoustic pressure.

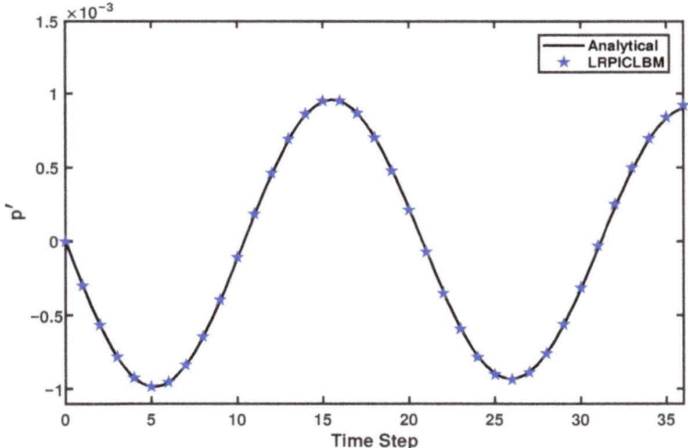

Figure 6. Acoustic pressure ($\frac{kg}{\Delta x \Delta t^2}$) versus time step ($\Delta t$) for $\nu = 1.0 \times 10^{-2}$ ($\frac{\Delta x^2}{\Delta t}$) with $N_{ppw} = 12$ points per wavelength with irregular nodal distributions.

3.2. Planar Pulse Wave

Next, we studied a planar pulse wave by replacing the plane wave with a Gaussian shape planar pulse, initially located at the center of the domain. The dissipation and dispersion relationships derived from the spatial analysis of Equation (26) are [25]

$$c_S = \sqrt{2}c_s \sqrt{\frac{1 + (\omega\tau_s)^2}{\sqrt{1 + (\omega\tau_s)^2} + 1}} \tag{29a}$$

$$\alpha_S = \frac{\omega}{\sqrt{2}c_s} \sqrt{\frac{\sqrt{1 + (\omega\tau_s)^2} - 1}{1 + (\omega\tau_s)^2}} \tag{29b}$$

The variables and parameters for this case are presented in Table 2. A planar pulse emits from the origin throughout the domain, where periodic boundary conditions are imposed.

Table 2. Parameters for the planar pulse wave.

Variables	$p'(x,y,0)$	ρ'	u'	v'	A	σ
Description	$A\exp\left(-\ln(2)\frac{x^2}{\sigma^2}\right)$	$\frac{p'}{c_s^2}$	$\frac{p'}{\rho_0 c_s}$	0	$10^{-3}p_0$	$0.06 - 0.1$

To assess the accuracy of the LRPIC-LBM in simulating the pulse wave emission, we proceeded as in reference [26]. Different viscosities and resolutions (which are related to the choice of σ) were studied. Figure 7 depicts the acoustic pressure time history for $\nu = 1.0 \times 10^{-2}$ ($\frac{\Delta x^2}{\Delta t}$) with $\sigma = 0.1$. The cumulant LBM results (dashed red line) are compared to the LRPIC-LBM (solid black line). As stated in [25,26] the intensity loss at any location is proportional to the distance of propagation, regardless of the precise location. Thus, the data were extracted from the center of the domain, and at 5, 11, and 17 nodes apart from the center. For resolutions up to $\sigma = 0.1$, deviations between the cumulant LBM and the LRPIC-LBM results were less than 1 percent.

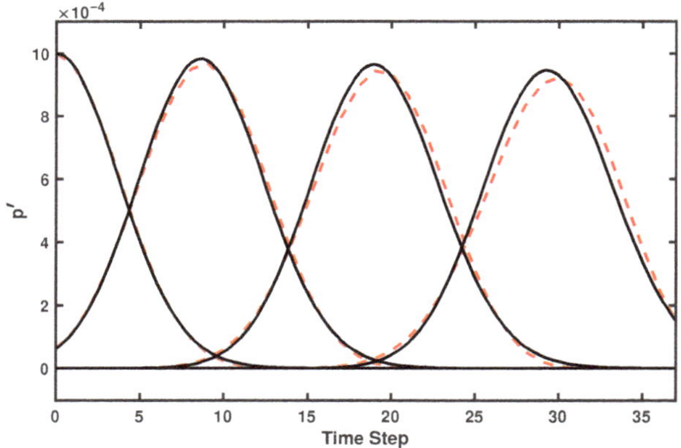

Figure 7. Acoustic pressure ($\frac{kg}{\Delta x \Delta t^2}$) versus time step ($\Delta t$) for $\nu = 1.0 \times 10^{-2}$ ($\frac{\Delta x^2}{\Delta t}$) with $\sigma = 0.1$: the cumulant LBM (dashed red line), LRPIC-LBM (solid black line).

The acoustic pressure time history for $\nu = 1.0 \times 10^{-4}$ ($\frac{\Delta x^2}{\Delta t}$) with $\sigma = 0.1$ is shown in Figure 8. It presents the comparison between the cumulant LBM and the LRPIC-LBM at low viscosity. While the standard BGK LBM creates instabilities and noisy results, less than 1 percent deviation was found between the cumulant LBM and this method, with stable behavior at low viscosities. In addition, as in the case of the temporal results shown in Figure 3, reducing the viscosity in the considered range did not substantially impact the results.

Like the parameter N_{ppw} introduced in the temporal analysis of the first set up, σ is another effective parameter used in spatial analysis which affects the accuracy of the LBM results. The acoustic pressure time history is depicted in Figure 9 for $\nu = 1.0 \times 10^{-2}$ ($\frac{\Delta x^2}{\Delta t}$), with $\sigma = 0.06$. It shows that the deviations between the cumulant LBM and the LRPIC-LBM results increase after the reduction of σ. The wiggling observed in the results of the cumulant LBM is due to the reduction of the number of nodes inside the pulse, which brings the accuracy of the results into question. However, the LRPIC-LBM graphs are smooth and unaffected by the lesser number of nodes. To better assess the accuracy of the LRPIC-LBM compared to the cumulant LBM, the data shown in Figure 9 were compared with the analytical results. The Fourier transform of the pressure time history of the passing wave yields the Fourier coefficient of pressure $\widehat{p'}(x, \omega) = \widehat{p'}(x, 0) \exp[-\alpha_S x] \exp\left[i\omega \frac{x}{c_S}\right]$ which is the solution to Equation (26) [25]. Figure 10 shows the ratio $\widehat{p'}(x, \omega)/\widehat{p'}(x, 0)$ as a function of angular frequency for the analytical solution. This figure shows that the LRPIC-LBM is more successful than the cumulant LBM at predicting theoretical results with $\sigma = 0.06$. Thus, this method can model wave propagation more accurately even at smaller resolutions.

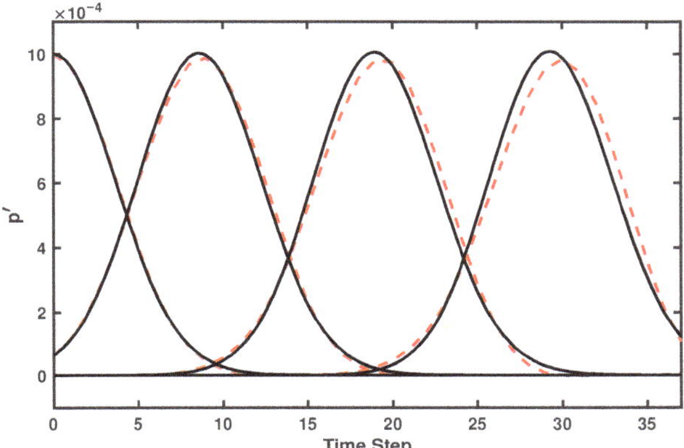

Figure 8. Acoustic pressure $(\frac{kg}{\Delta x \Delta t^2})$ versus time step (Δt) for $\nu = 1.0 \times 10^{-4}$ $(\frac{\Delta x^2}{\Delta t})$ with $\sigma = 0.1$: the cumulant LBM (dashed red line), LRPIC-LBM (solid black line).

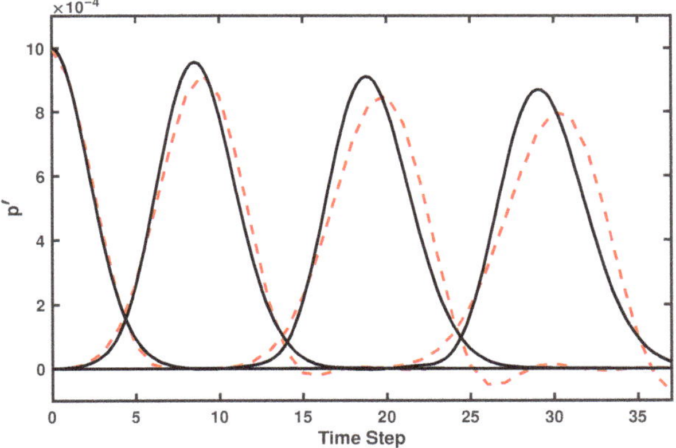

Figure 9. Acoustic pressure$(\frac{kg}{\Delta x \Delta t^2})$ versus time step (Δt) for $\nu = 1.0 \times 10^{-2}$ $(\frac{\Delta x^2}{\Delta t})$ with $\sigma = 0.06$: the cumulant LBM (dashed red line), LRPIC-LBM (solid black line).

Figure 11 illustrates the acoustic pressure time history for $\nu = 1.0 \times 10^{-2}$ $(\frac{\Delta x^2}{\Delta t})$ with $\sigma = 0.1$ considering irregular nodes (Figure 1b). The cumulant LBM and the LRPIC-LBM results are represented by a dashed blue line and a solid black line, respectively. The data are extracted as in Figure 7. The results show that the LRPIC-LBM with irregular nodal distributions reproduced the behavior observed previously.

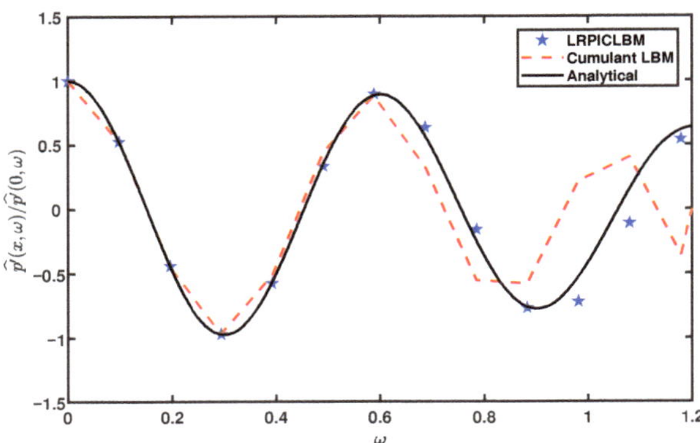

Figure 10. The ratio $\widehat{p'}(x,\omega)/\widehat{p'}(x,0)$ as a function of angular frequency $(\frac{1}{\Delta t})$ for the analytical solution, the cumulant LB and the LRPIC-LB methods.

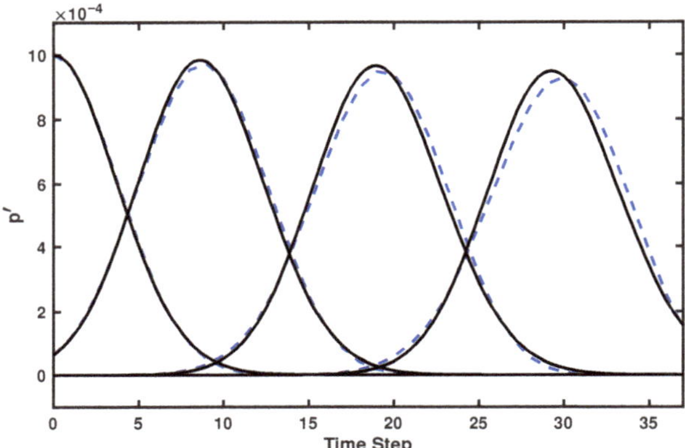

Figure 11. Acoustic pressure $(\frac{kg}{\Delta x \Delta t^2})$ versus time step (Δt) for $\nu = 1.0 \times 10^{-2}$ $(\frac{\Delta x^2}{\Delta t})$ with $\sigma = 0.1$ considering irregular nodal distributions: the cumulant LBM (dashed blue line), LRPIC-LBM (solid black line).

4. Conclusions

This paper presents a numerical study of the propagation of acoustic waves, one of the challenging issues occurring in wind turbine simulations, that includes the temporal decay of a standing plane wave and the spatial decay of a planar acoustic pulse, using the local radial point interpolation cumulant lattice Boltzmann method (LRPIC-LBM). The LB equation is divided into collision, modeled by the cumulant method, and streaming, discretized in time and space, using the Lax–Wendroff scheme and the local radial point interpolation method (RPIM). The LRPIC-LBM results were compared with the cumulant LBM and the analytical solutions. Both methods showed a similar acoustic pressure time history, and the deviations for the phase speed and the dissipation rate were minor for high number of points per wavelength N_{ppw} and resolution σ. In addition, they showed that reduced viscosity does not affect the stability of either LB method due to the intrinsic characteristics of the cumulant method. Unlike LB methods such as the BGK LBM and the cumulant LBM, the time history for the acoustic pressure and the phase speed and dissipation rate predicted by LRPIC-LBM showed considerably smaller errors for low N_{ppw} and σ due to the construction of the meshless method itself. Moreover, the LRPIC-LBM with irregular nodal distributions reproduces the same propagation of acoustic waves obtained with regular nodal distributions.

In summary, the freedom to scatter nodes based on problem conditions and the occurrence of sharp gradients plus the accuracy obtained from the RPIM along with the good stability and simplicity achieved by the cumulant LBM may provide an adequate platform with which to model wind turbine problems.

Author Contributions: Conceptualization: M.G., C.S., I.C. and E.K.F.; methodology, M.G., C.S., I.C. and E.K.F.; coding, M.G., C.S., I.C. and E.K.F.; validation, M.G., C.S., I.C. and E.K.F.; formal analysis, M.G., C.S., I.C. and E.K.F.; investigation, M.G., C.S., I.C. and E.K.F.; resources, M.G., C.S., I.C. and E.K.F.; data curation, M.G., C.S., I.C. and E.K.F.; writing—original draft preparation, M.G.; writing—review and editing, C.S., I.C. and E.K.F.; visualization, M.G., C.S., I.C. and E.K.F.; supervision, C.S. and I.C.; project administration, C.S. and I.C.; funding acquisition, C.S. and I.C. All authors contributed equally to this work. All authors have read and agreed to the published version of the manuscript.

Funding: This work has been funded by the Spanish Ministry of Economy, Industry and Competitiveness Research National Agency (under project DPI2016-75791-C2-1-P), by FEDER funds and by Generalitat de Catalunya - AGAUR (under project 2017 SGR 01234).

Institutional Review Board Statement: Not applicable.

Informed Consent Statement: Not applicable.

Data Availability Statement: Not applicable.

Conflicts of Interest: The authors declare no conflict of interest.

Abbreviations

The following abbreviations are used in this manuscript:

LRPIC-LBM	Local radial point interpolation cumulant lattice Boltzmann method
LBM	Lattice Boltzmann method
RPIM	Radial point interpolation method
NREL	National Renewable Energy Laboratory
NWTC	National Wind Technology Center
CAA	Computational aeroacoustic
DRP	Dispersion relation preserving scheme
GODRP	Grid optimized dispersion relation preserving schemes
BGK-LBM	Bhatnagar Gross Krook lattice Boltzmann method
LBM-rrBGK	Lattice Boltzmann method recursive regularized Bhatnagar Gross Krook
IS-LBM	Interpolation supplemented lattice Boltzmann method
FDLBE	Finite difference lattice Boltzmann equation
FVLBE	Finite volume lattice Boltzmann Equation
FELBE	Finite element lattice Boltzmann equation
TLLBM	Taylor-series expansion and least-squares-based lattice Boltzmann method
FSI	Fluid structure interaction
FDM	Finite difference method
SPH	Smoothed particle hydrodynamics
LPIM	Local point interpolation method
LRPIM	Local radial point interpolation method
MLPG	Meshless local Petrov–Galerkin method
MRT	Multi-relaxation times
PDF	Probability density function
PIM	Point interpolation method
RBF	Radial basis functions
MQ	Multi-quadrics
WR	Weighted residuals
PDE	Partial differential equation

References

1. Burton-Jones, A. *Knowledge Capitalism: Business, Work, and Learning in the New Economy*; Oxford University Press: Oxford, UK, 2001.
2. Letcher, T.M. *Wind Energy Engineering: A Handbook for Onshore and Offshore Wind Turbines*; Academic Press: Cambridge, MA, USA, 2017.
3. Bowdler, D.; Leventhall, H. *Wind Turbine Noise*; Multi-Science Pub.: Brentwood, CA, USA, 2011.
4. Wagner, S.; Bareiss, R.; Guidati, G. *Wind Turbine Noise*; Springer Science & Business Media: Berli/Heidelberg, Germany, 2012
5. Singer, B.A.; Lockard, D.P.; Brentner, K.S. Computational aeroacoustic analysis of slat trailing-edge flow. *AIAA J.* **2000**, *38*, 1558–1564.
6. Manwell, J.F.; McGowan, J.G.; Rogers, A.L. *Wind Energy Explained: Theory, Design and Application*; John Wiley & Sons: Hoboken, NJ, USA, 2010.
7. Geier, M.; Kian Far, E. A Sliding Grid Method For The Lattice Boltzmann Method Using Compact Interpolation. In Proceedings of the The First International Workshop on Lattice Boltzmann for Wind Energy, Online Conference, 25–26 February 2021.
8. Morris, P.; Long, L.; Brentner, K. An aeroacoustic analysis of wind turbines. In Proceedings of the 42nd AIAA Aerospace Sciences Meeting and Exhibit, Reno, NV, USA, 5–8 January 2004; p. 1184.
9. Robinson, M.C.; Hand, M.; Simms, D.; Schreck, S. *Horizontal Axis Wind Turbine Aerodynamics: Three-Dimensional, Unsteady, and Separated Flow Influences*; Technical Report; National Renewable Energy Lab.: Golden, CO, USA, 1999.
10. Tummala, A.; Velamati, R.K.; Sinha, D.K.; Indraja, V.; Krishna, V.H. A review on small scale wind turbines. *Renew. Sustain. Energy Rev.* **2016**, *56*, 1351–1371.
11. Spera, D.A. *Wind Turbine Technology*; Wind Energy Technologies Office: Washington, DC, USA, 1994.
12. Tam, C.K. Computational aeroacoustics-Issues and methods. *AIAA J.* **1995**, *33*, 1788–1796.
13. Wells, V.L.; Renaut, R.A. Computing aerodynamically generated noise. *Annu. Rev. Fluid Mech.* **1997**, *29*, 161–199.
14. Kim, J.W.; Lee, D.J. Optimized compact finite difference schemes with maximum resolution. *AIAA J.* **1996**, *34*, 887–893.
15. Tam, C.K.; Webb, J.C. *Dispersion-Relation-Preserving Schemes for Computational Aeroacoustics*; NASA: Washington, DC, USA, 1992.
16. Cheong, C.; Lee, S. Grid-optimized dispersion-relation-preserving schemes on general geometries for computational aeroacoustics. *J. Comput. Phys.* **2001**, *174*, 248–276.
17. Gröschel, E.; Schröder, W.; Renze, P.; Meinke, M.; Comte, P. Noise prediction for a turbulent jet using different hybrid methods. *Comput. Fluids* **2008**, *37*, 414–426.
18. Delfs, J.; Bertsch, L.; Zellmann, C.; Rossian, L.; Kian Far, E.; Ring, T.; Langer, S.C. Aircraft Noise Assessment—From Single Components to Large Scenarios. *Energies* **2018**, *11*, 429.
19. Colonius, T.; Lele, S.K.; Moin, P. The scattering of sound waves by a vortex: Numerical simulations and analytical solutions. *J. Fluid Mech.* **1994**, *260*, 271–298.
20. Mitchell, B.; Lele, S.; Moin, P. Direct computation of the sound from a compressible co-rotating vortex pair. In Proceedings of the 30th Aerospace Sciences Meeting and Exhibit, Reno, NV, USA, 6–9 January 1992; p. 374.
21. Kian Far, E.; Geier, M.; Kutscher, K.; Krafczyk, M. Implicit Large Eddy Simulation of Flow in a Micro-Orifice with the Cumulant Lattice Boltzmann Method. *Computation* **2017**, *5*, 23, doi:10.3390/computation5020023.
22. Kian Far, E. A Cumulant LBM approach for Large Eddy Simulation of Dispersion Microsystems. Ph.D. Thesis, Universität Braunschweig, Braunschweig, Germany, 2015.
23. Buick, J.; Greated, C.; Campbell, D. Lattice BGK simulation of sound waves. *EPL (Europhysics Lett.)* **1998**, *43*, 235.
24. Dellar, P.J. Bulk and shear viscosities in lattice Boltzmann equations. *Phys. Rev. E* **2001**, *64*, 031203.
25. Bres, G.; Pérot, F.; Freed, D. Properties of the lattice Boltzmann method for acoustics. In Proceedings of the 15th AIAA/CEAS Aeroacoustics Conference (30th AIAA Aeroacoustics Conference), Miami, FL, USA, 11–13 May 2009; p. 3395.
26. Gorakifard, M.; Cuesta, I.; Salueña, C.; Kian Far, E. Acoustic Wave Propagation and its Application to Fluid Structure Interaction using the Cumulant Lattice Boltzmann Method. *Comput. Math. Appl.* **2021**, *87*, 91–106.
27. Latt, J.; Chopard, B. Lattice Boltzmann method with regularized pre-collision distribution functions. *Math. Comput. Simul.* **2006**, *72*, 165–168.
28. Malaspinas, O. Increasing stability and accuracy of the lattice Boltzmann scheme: recursivity and regularization. *arXiv* **2015**, arXiv:1505.06900.
29. Brogi, F.; Malaspinas, O.; Chopard, B.; Bonadonna, C. Hermite regularization of the lattice Boltzmann method for open source computational aeroacoustics. *J. Acoust. Soc. Am.* **2017**, *142*, 2332–2345.
30. Chávez-Modena, M.; Martínez, J.; Cabello, J.; Ferrer, E. Simulations of aerodynamic separated flows using the lattice Boltzmann solver XFlow. *Energies* **2020**, *13*, 5146.
31. Filippova, O.; Hänel, D. Grid Refinement for Lattice-BGK Models. *J. Comput. Phys.* **1998**, *147*, 219–228, doi:10.1006/jcph.1998.6089.
32. Wood, S.L. Lattice Boltzmann Methods for Wind Energy Analysis. Ph.D. Thesis, University of Tennessee, Knoxville, TN, USA, 2016.
33. Deiterding, R.; Wood, S.L. *Predictive Wind Turbine Simulation with an Adaptive Lattice Boltzmann Method for Moving Boundaries*; Journal of Physics: Conference Series; IOP Publishing: Bristol, UK, 2016; Volume 753, p. 082005.

34. Gorakifard, M.; Salueña, C.; Cuesta, I.; Kian Far, E. Acoustical analysis of fluid structure interaction using the Cumulant lattice Boltzmann method. In Proceedings of the 16th International Conference for Mesoscopic Methods in Engineering and Science, Heriot-Watt University, Edinburgh, UK, 22–26 July 2019.
35. Chen, Z.; Shu, C.; Tan, D.S. The simplified lattice Boltzmann method on non-uniform meshes. *Commun. Comput. Phys* **2018**, *23*, 1131.
36. He, X.; Luo, L.S.; Dembo, M. Some Progress in Lattice Boltzmann Method. Part I. Nonuniform Mesh Grids. *J. Comput. Phys.* **1996**, *129*, 357 – 363, doi:10.1006/jcph.1996.0255.
37. He, X.; Doolen, G. Lattice Boltzmann Method on Curvilinear Coordinates System: Flow around a Circular Cylinder. *J. Comput. Phys.* **1997**, *134*, 306 – 315, doi:10.1006/jcph.1997.5709.
38. Chen, H. Volumetric formulation of the lattice Boltzmann method for fluid dynamics: Basic concept. *Phys. Rev. E* **1998**, *58*, 3955–3963, doi:10.1103/PhysRevE.58.3955.
39. Mei, R.; Shyy, W. On the finite difference-based lattice Boltzmann method in curvilinear coordinates. *J. Comput. Phys.* **1998**, *143*, 426–448.
40. Xi, H.; Peng, G.; Chou, S.H. Finite-volume lattice Boltzmann method. *Phys. Rev. E* **1999**, *59*, 6202.
41. Nannelli, F.; Succi, S. The lattice Boltzmann equation on irregular lattices. *J. Stat. Phys.* **1992**, *68*, 401–407.
42. Peng, G.; Xi, H.; Duncan, C.; Chou, S.H. Finite volume scheme for the lattice Boltzmann method on unstructured meshes. *Phys. Rev. E* **1999**, *59*, 4675.
43. Lee, T.; Lin, C.L. A Characteristic Galerkin Method for Discrete Boltzmann Equation. *J. Comput. Phys.* **2001**, *171*, 336–356, doi:10.1006/jcph.2001.6791.
44. Li, Y.; LeBoeuf, E.J.; Basu, P.K. Least-squares finite-element scheme for the lattice Boltzmann method on an unstructured mesh. *Phys. Rev. E* **2005**, *72*, 046711, doi:10.1103/PhysRevE.72.046711.
45. Min, M.; Lee, T. A spectral-element discontinuous Galerkin lattice Boltzmann method for nearly incompressible flows. *J. Comput. Phys.* **2011**, *230*, 245–259, doi:10.1016/j.jcp.2010.09.024.
46. Shu, C.; Niu, X.; Chew, Y. Taylor-series expansion and least-squares-based lattice Boltzmann method: Two-dimensional formulation and its applications. *Phys. Rev. E* **2002**, *65*, 036708.
47. Shu, C.; Chew, Y.; Niu, X. Least-squares-based lattice Boltzmann method: A meshless approach for simulation of flows with complex geometry. *Phys. Rev. E* **2001**, *64*, 045701.
48. Fard, E.G.; Shirani, E.; Geller, S. The Fluid Structure Interaction Using the Lattice Boltzmann Method. In Proceedings of the 13th Annual International Conference fluid dynamic conference, Shiraz, Iran, 26–28 October 2010.
49. Liu, G.R.; Gu, Y.T. *An Introduction to Meshfree Methods and Their Programming*; Springer Science & Business Media: Berlin/Heidelberg, Germany, 2005.
50. Liu, G. *Meshfree Methods: Moving Beyond the Finite Element Method, Second Edition*, 2nd ed.; CRC Press: Boca Raton, FL, USA, 2009; pp. 1–692.
51. Liu, G.; Gu, Y. A local point interpolation method for stress analysis of two-dimensional solids. *Struct. Eng. Mech.* **2001**, *11*, 221–236.
52. Liu, G.; Gu, Y. A local radial point interpolation method (LRPIM) for free vibration analyses of 2-D solids. *J. Sound Vib.* **2001**, *246*, 29–46.
53. Atluri, S.N.; Zhu, T. A new meshless local Petrov-Galerkin (MLPG) approach in computational mechanics. *Comput. Mech.* **1998**, *22*, 117–127.
54. Li, Q.; Yang, H.; Yang, F.; Yao, D.; Zhang, G.; Ran, J.; Gao, B. Calculation of Hybrid Ionized Field of AC/DC Transmission Lines by the Meshless Local Petorv–Galerkin Method. *Energies* **2018**, *11*, 1521.
55. Musavi, S.H.; Ashrafizaadeh, M. Meshless lattice Boltzmann method for the simulation of fluid flows. *Phys. Rev. E* **2015**, *91*, 023310.
56. Musavi, S.H.; Ashrafizaadeh, M. Development of a three dimensional meshless numerical method for the solution of the Boltzmann equation on complex geometries. *Comput. Fluids* **2019**, *181*, 236–247.
57. Bawazeer, S. Lattice Boltzmann Method with Improved Radial Basis Function Method. Ph.D. Thesis, University of Calgary, Calgary, AB, Canada, 2019.
58. Tanwar, S. A meshfree-based lattice Boltzmann approach for simulation of fluid flows within complex geometries: Application of meshfree methods for LBM simulations. In *Analysis and Applications of Lattice Boltzmann Simulations*; IGI Global: Hershey, PA, USA, 2018; pp. 188–222.
59. Kian Far, E.; Geier, M.; Krafczyk, M. Simulation of rotating objects in fluids with the cumulant lattice Boltzmann model on sliding meshes. *Comput. Math. Appl.* **2018**, doi:10.1016/j.camwa.2018.08.055.
60. Kian Far, E.; Geier, M.; Kutscher, K.; Konstantin, M. Simulation of micro aggregate breakage in turbulent flows by the cumulant lattice Boltzmann method. *Comput. Fluids* **2016**, *140*, 222 – 231, doi:10.1016/j.compfluid.2016.10.001.
61. Kian Far, E.; Langer, S. Analysis of the cumulant lattice Boltzmann method for acoustics problems. In Proceedings of the 13th International Conference on Theoretical and Computational Acoustics, Vienna, Austria, 31 July–3 August 2017.
62. Geier, M. Ab Initio Derivation of the Cascaded Lattice Boltzmann Automaton. Ph.D. Thesis, University of Freiburg–IMTEK, Freiburg, Germany, 2006.

63. Geier, M.; Greiner, A.; Korvink, J.G. *Reference Frame Independent Partitioning of the Momentum Distribution Function in Lattice Boltzmann Methods with Multiple Relaxation Rates*; University of Freiburg: Freiburg, Germany, 2008.
64. Seeger, S.; Hoffmann, H. The cumulant method for computational kinetic theory. *Contin. Mech. Thermodyn.* **2000**, *12*, 403–421.
65. Seeger, S.; Hoffmann, K. The cumulant method for the space-homogeneous Boltzmann equation. *Contin. Mech. Thermodyn.* **2005**, *17*, 51–60.
66. Kian Far, E.; Geier, M.; Kutscher, K.; Krafczyk, M. Distributed cumulant lattice Boltzmann simulation of the dispersion process of ceramic agglomerates. *J. Comput. Methods Sci. Eng.* **2016**, *16*, 231–252.
67. Kian Far, E. A sliding mesh LBM approach for the simulation of the rotating objects. In Proceedings of the 13th International Conference for Mesoscopic Methods in Engineering and Science, Hamburg, Germany, 18–22 July 2016.
68. Geier, M.; Schönherr, M.; Pasquali, A.; Krafczyk, M. The cumulant lattice Boltzmann equation in three dimensions: Theory and validation. *Comput. Math. Appl.* **2015**, *70*, 507–547.
69. Kian Far, E. Turbulent Flow Simulation of Dispersion Microsystem with Cumulant Lattice Boltzmann Method, FORMULA X, 24 June 2019. Available online: https://www.formulation.org.uk/images/stories/FormulaX/Posters/P-27.pdf (accessed on 2 March 2021).
70. Schäfer, R.; Böhle, M. Validation of the Lattice Boltzmann Method for Simulation of Aerodynamics and Aeroacoustics in a Centrifugal Fan. In *Acoustics*; Multidisciplinary Digital Publishing Institute: Basel, Switzerland, 2020; Volume 2, pp. 735–752.
71. Ye, Q.; Avallone, F.; Van Der Velden, W.; Casalino, D. *Effect of Vortex Generators on NREL Wind Turbine: Aerodynamic Performance and Far-Field Noise*; Journal of Physics: Conference Series; IOP Publishing: Bristol, UK, 2020; Volume 1618, p. 052077.
72. Kinsler, L.E.; Frey, A.R.; Coppens, A.B.; Sanders, J.V. *Fundamentals of Acoustics*; John Wiley & Sons: Hoboken, NJ, USA, 1999.

Article

Numeric Simulation of Acoustic-Logging of Cave Formations

Fanghui Xu and Zhuwen Wang *

College of Geoexploration Science and Technology, Jilin University, Changchun 130026, China; xufh19@mails.jlu.edu.cn
* Correspondence: wangzw@jlu.edu.cn; Tel.: +86-135-7894-7784

Received: 1 July 2020; Accepted: 22 July 2020; Published: 31 July 2020

Abstract: The finite difference (FD) method of monopole source is used to simulate the response of full-wave acoustic-logging in cave formations. The effect of the cave in the formation of borehole full-waves was studied. The results show that the radius of cave is not only linearly related to the first arrival of the compressional wave (P-wave), but also to the energy of the shear wave (S-wave). The converted S (S–S wave) and P-waves (S–P wave) are formed when the S-wave encounters the cave. If the source distance is small, the S–S and S–P waves are not separated, and the attenuation of the S-wave is not large, due to superposition of the converted waves. The S–P wave has been separated from the S-wave when the source distance is large, so the attenuation of the S-wave increases. The amplitude of the P and S–waves changes most when the distance of the cave to the borehole wall reaches a certain value; this value is related to the excitation frequency. The amplitude of the Stoneley wave (ST wave) varies directly with the radius of cave. If the radius of the cave is large, the energy of ST wave is weak. The scattered wave is determined by the radius and position of the cave. The investigation depth of a monopole source is limited. When the distance of the cave to the borehole wall exceeds the maximum investigation depth, the borehole acoustic wave is little affected by the cave. In actual logging, the development of the cave can be evaluated by using the first arrival of the P-wave and the energy of the S and ST waves.

Keywords: cave formation; P-waves; S-waves; Stoneley wave; scattered wave

1. Introduction

Caves are widely distributed in carbonate and igneous reservoirs, causing heterogeneity and making it difficult to evaluate the reservoirs by conventional logging. Based on the observation of cores, the reservoir spaces of igneous and carbonate formations can be divided into primary pores and secondary pores, according to their genesis [1]. Secondary pores contain caves. These can be divided into matrix-dissolution pores, intragranular-dissolution pores and apricot-kernel dissolution pores in igneous formations [2]. According to the genesis, carbonate formations have porous caves, intergranular caves and fractured caves. On the basis of the size of the cave, they may fall into middle and small-type caves and large caves. Generally, caves with a diameter greater than 10 mm can be regarded as large caves [3]. Because dissolution pores and caves often transform primary pores, making them better connected and improving the reservoir property of the rock, they are important reservoir spaces [4]. In fact, the caves in the formation can be defined as voids, washouts or large pores of secondary porosity in a more general. The existence of caves greatly changes the speed and energy of acoustic-logging [5]. Therefore, it is very important to grasp the effect of caves on borehole acoustic wave propagation, which plays an important role in the detection and evaluation of caves. Currently, the effect of fractured and gas-bearing reservoirs on borehole acoustics has been investigated [6–8]. However, there are few studies on the acoustic-logging response of caves. Heterogeneous bodies such as fractures and caves, scatter seismic-wave propagation [9]. Elastic-wave scattering in heterogeneous media can be divided into "velocity" and "wave impedance" types [10]. Due to the fact that the scattered wave of a borehole dipole source is mainly a SH shear wave [11–13], the dipole source has been used to study the scattering effect in heterogeneous formation [14]. The relationship between S-wave wavelength and the size of cave can be obtained by dipole acoustic-logging [15]. According to the different filling materials, caves can be divided into four types: unfilled, sand-mud-filled, breccia filled, and calcite-filled [16]. The filling degree of a cave can slow P and S-waves. Studies have shown that different filling degrees of caves has a good corresponding relationship with acoustic simulation curves, and the results have certain application effect [17]. In cave formation, the slowness of the P-wave increases and the density of well-logging decreases. The resistivity is generally low in conventional logging [18]. Acoustic-logging and electrical-imaging logging can evaluate caves comprehensively; ST wave energy of acoustic-logging reduces due to caves [3].

The above studies show that it is feasible to use acoustic-logging to evaluate near-borehole caves. Based on actual logging data, the results have qualitatively shown the effects of cave on conventional logging, such as P-wave slowness. The response of component waves to caves has been studied. There is no detailed regularity of the size and position of cave on borehole acoustic waves. Monopole is a common source in acoustic-logging [19,20], and the component waves with different frequencies formed by monopole contain the information of specific formations. It is difficult to analyze scattered-wave fields of monopole acoustic-logging because of their complexity. Numeric simulation in cave reservoirs by monopole sources has not been reported. In this study, the finite difference (FD) method of the elastic wave equation is used to simulate a monopole full-wave acoustic-logging response of a fluid-filled cave in a homogeneous formation [21,22]. On this basis, the results of a numeric simulation are used to analyze actual logging data [23].

In the process of simulating a wave field inside and outside of a borehole by FD, matrix calculations are mainly carried out. It is necessary to divide small grids to meet the calculation requirements of cave, which greatly increases the time costs of the numeric simulation. Therefore, computer-unified device architecture (CUDA) parallel-computing was used to improve the calculation efficiency [24–27]. The essence of CUDA is to change the computing mode from central processing unit (CPU) serial to graphics-processing unit (GPU) parallel.

2. Theory and Methods

2.1. FD Method for Elastic Wave Equation

In order to study the issue more simply, we use the elastic wave equation in the 2D-cylindrical coordinate system. A 2D-model simulation has advantages over a 3D-model simulation in terms of calculation times. A 2D calculation of a cave can also show the main characteristics of the problem studied [15]. The staggered-grid finite difference method is used to simulate the propagation of elastic waves in our simulation [6]. The wave equation of velocity stress in isotropic media in cylindrical coordinates is as follows:

$$\left[\tau^{n+1}_{rr(i+\frac{1}{2},j+\frac{1}{2})} - \tau^{n}_{rr(i+\frac{1}{2},j+\frac{1}{2})}\right]/\Delta t = \left[\lambda_{(i+\frac{1}{2},j+\frac{1}{2})} + 2\mu_{i+\frac{1}{2},j+\frac{1}{2}}\right]\left[L^{1}_{R}v^{n+\frac{1}{2}}_{r(i,j+\frac{1}{2})}\right] + \lambda_{(i+\frac{1}{2},j+\frac{1}{2})}\left[L^{1}_{Z}v^{n+\frac{1}{2}}_{z(i+\frac{1}{2},j)} + \sigma_i v^{n+\frac{1}{2}}_{r(i,j+\frac{1}{2})}/r_{i+\frac{1}{2}}\right] \tag{1}$$

$$\left[\tau^{n+1}_{zz(i+\frac{1}{2},j+\frac{1}{2})} - \tau^{n}_{zz(i+\frac{1}{2},j+\frac{1}{2})}\right]/\Delta t = \lambda_{(i+\frac{1}{2},j+\frac{1}{2})}\left[L^{1}_{R}v^{n+\frac{1}{2}}_{r(i,j+\frac{1}{2})} + \frac{\sigma_i v^{n+\frac{1}{2}}_{r(i,j+\frac{1}{2})}}{r_{i+\frac{1}{2}}}\right] + \left[\lambda_{(i+\frac{1}{2},j+\frac{1}{2})} + 2\mu_{i+\frac{1}{2},j+\frac{1}{2}}\right]\left[L^{1}_{Z}v^{n+\frac{1}{2}}_{z(i+\frac{1}{2},j)}\right] \tag{2}$$

$$\left[\tau^{n+1}_{\theta\theta(i+\frac{1}{2},j+\frac{1}{2})} - \tau^{n}_{\theta\theta(i+\frac{1}{2},j+\frac{1}{2})}\right]/\Delta t = \lambda_{(i+\frac{1}{2},j+\frac{1}{2})}\left[L^{1}_{R}v^{n+\frac{1}{2}}_{r(i,j+\frac{1}{2})} + L^{1}_{Z}v^{n+\frac{1}{2}}_{z(i+\frac{1}{2},j)}\right] + \left[\lambda_{(i+\frac{1}{2},j+\frac{1}{2})} + 2\mu_{i+\frac{1}{2},j+\frac{1}{2}}\right]\left[\sigma_i v^{n+\frac{1}{2}}_{r(i,j+\frac{1}{2})}/r_{i+\frac{1}{2}}\right] \tag{3}$$

$$\left[\tau^{n+1}_{rz(i,j)} - \tau^{n}_{rz(i,j)}\right]/\Delta t = \mu^{H}_{(i-\frac{1}{2},j-\frac{1}{2})}\left[L^{1}_{R}v^{n+\frac{1}{2}}_{z(i-\frac{1}{2},j)} + L^{1}_{Z}v^{n+\frac{1}{2}}_{r(i,j-\frac{1}{2})}\right] \tag{4}$$

$$\sigma_i \rho_{(i-\frac{1}{2},j+\frac{1}{2})}\left[v^{n+\frac{1}{2}}_{r(i,j+\frac{1}{2})} - v^{n-\frac{1}{2}}_{r(i,j+\frac{1}{2})}\right]/\Delta t = L^1_R \tau^n_{rr(i-\frac{1}{2},j+\frac{1}{2})} + L^1_Z \tau^n_{rz(i,j)}$$
$$+ \left[\sigma_i \tau^n_{rr(i-\frac{1}{2},j+\frac{1}{2})} - \sigma_i \tau^n_{\theta\theta(i-\frac{1}{2},j+\frac{1}{2})}\right]/r_i \tag{5}$$

$$\sigma_i \rho_{(i+\frac{1}{2},j-\frac{1}{2})}\left[v^{n+\frac{1}{2}}_{z(i+\frac{1}{2},j)} - v^{n-\frac{1}{2}}_{z(i+\frac{1}{2},j)}\right]/\Delta t = L^1_R \tau^n_{xz(i,j)} + L^1_Z \tau^n_{zz(i+\frac{1}{2},j-\frac{1}{2})} + \sigma_i \tau^n_{rz(i,j)}/r_{i+\frac{1}{2}} \tag{6}$$

In the equations mentioned above, the subscripts i and j in brackets refer to spatial index of nodes in x and z directions, respectively, the superscript n represents the time discrete index, τ_{rr}, $\tau_{\theta\theta}$ and τ_{zz} refer to normal stress in x, θ and z directions, τ_{rz} is shear stress, Δt is the time step, L^1_R and L^1_Z are second-order difference operators of space, and the parameters ρ and μ represent the density and shear moduli of the medium. On the medium interface, the arithmetic mean value of ρ and the harmonic mean value of μ are employed; on the solid–liquid interface, μ is set to 0 [6,7]. $\sigma_i f_{(i,j)}$ and $\sigma_j f_{(i,j)}$ are arithmetic average operators and defined as:

$$\sigma_i f_{(i,j)} = 0.5\left[f_{(i+1,j)} + f_{(i,j)}\right] \tag{7}$$

$$\sigma_j f_{(i,j)} = 0.5\left[f_{(i,j)} + f_{(i,j+1)}\right] \tag{8}$$

$\mu^H_{(i,j)}$ is the harmonic mean and defined as

$$\mu^H_{(i,j)} = 4/\left[1/\mu_{(i+\frac{1}{2},j+\frac{1}{2})} + 1/\mu_{(i+\frac{1}{2},j-\frac{1}{2})} + 1/\mu_{(i-\frac{1}{2},j+\frac{1}{2})} + 1/\mu_{(i-\frac{1}{2},j-\frac{1}{2})}\right] \tag{9}$$

To reduce the numeric dispersion, the space grid needs to satisfy the following condition:

$$\Delta r \leq v_{min}/10 f_{max} \tag{10}$$

where v_{min} is the minimum velocity of numeric model, and f_{max} is the maximum frequency of the source [6,7]. The stability condition of the second-order FD is

$$v_{max}\Delta t\left[\frac{1}{\Delta r^2} + \frac{1}{\Delta z^2}\right]^{\frac{1}{2}} < 1 \tag{11}$$

where v_{max} is the maximum velocity of the numeric model [6,7]. The monopole source function is

$$s(t) = \begin{cases} \frac{1}{2}\left[1 + \cos\frac{2\pi}{T_c}\left(t - \frac{T_c}{2}\right)\right]\cos 2\pi f_0\left(t - \frac{T_0}{2}\right), 0 \leq t \leq T_c \\ 0, t > T_c \end{cases} \tag{12}$$

where f_0 denotes the center frequency of the source, t represents the time, and T_c is the acoustic source pulse width, $T_c = 2/f_0$.

2.2. The Principle of CUDA Parallel-Computing

The CUDA parallel-computing model of FD method of elastic wave equation is established; the algorithm implementation flow is shown in Figure 1. During the initialization of the calculation model, the CPU first allocates storage space in the GPU memory for the velocities, stresses and medium parameters. The kernel functions of stress, velocity fields and source are then designed. The GPU is controlled by the CPU to complete the step-by-step calculations of velocity and stress fields. It was found that for the homogeneous medium, the calculation time of the conventional program was 10,571.5 s; the calculation time of the parallel program was only 704.7 s through the simulation experiment. CUDA parallel-computing is used as a computing method in this study, so it is not introduced in detail here.

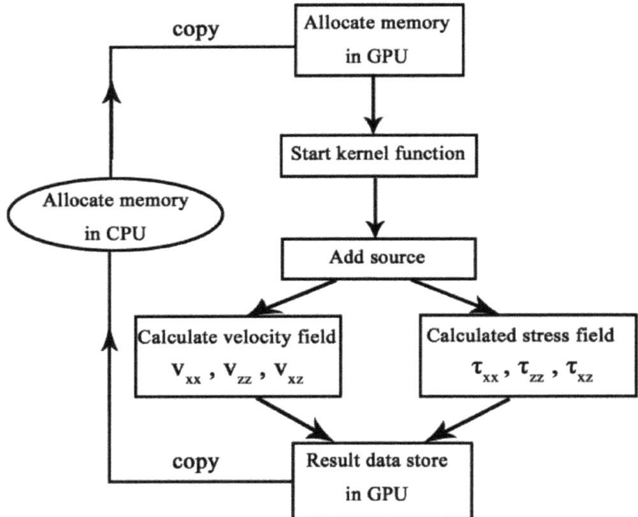

Figure 1. Graphics-processing unit (GPU) parallel-computing progress based on computer unified device architecture (CUDA) platform.

2.3. Borehole Model and Non-Splitting Perfect-Matched Layer Absorbing Boundary Condition

Figure 2 shows a borehole model surrounded by an elastic formation. The cave was filled with water and the formation was homogeneous. Perfect-matched layers (PML) were added at the top, bottom and right side of the model on the r_z plane. The model dimension was 4 m × 2 m, the grid step was $\Delta r = \Delta z = 2$ mm, and the time step was $\Delta t = 0.2$ μs. A non-splitting perfect-matched layer (NPML) approach [28] was used to mitigate the ghost reflection at the artificial boundary of the model.

The elastic parameters of each medium are shown in Table 1. Alteration usually occurs in carbonate and igneous reservoirs. The rock density of this kind of reservoir is generally between 2.5–2.8 g/cm^3. In this study, 2.65 g/cm^3 was selected as the density of the formation [20]. The P and S-wave velocity were selected according to the principle that the P-wave velocity is generally 1.6–1.9 times that of the S-wave velocity [20]. The selection of these physical parameters is similar in numeric simulations of fractures and cave formations [7,15,29]. If the physical parameters of calcite and dolomite are selected in the numeric simulation, it will be of more practical significance.

Table 1. Physical parameters of different media.

Elastic Parameters	v_p m/s	v_s m/s	ρ kg/m^3
Elastic formation	3670	2170	2650
Borehole fluid	1500	0	1000
Cave fluid	1500	0	1000

In order to verify the correctness of the results of the FD method, the propagation of the ST wave in homogeneous elastic formation was simulated by using the FD and the RAI method [29]. The formation parameters are shown in Table 1. The center frequency of the source was 2 kHz. Figure 3 shows the comparison between the waveform calculated by the FD and RAI method. The waveforms of the two methods matched well and it showed that the FD method was effective. In addition, there was a little difference in waveform amplitude between the two methods, which was caused by discrete Fourier transform (DFT) in RAI method and finite difference dispersion in FD method. The maximum error of the two methods was 1.2%.

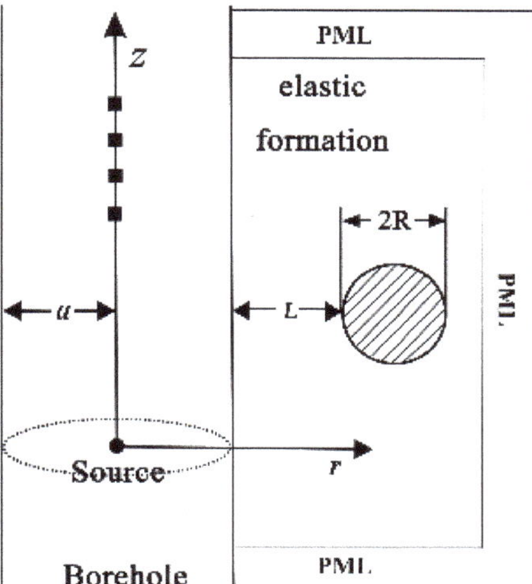

Figure 2. Schematic diagram of a cave formation in a borehole. The radius of the cave is *R*, the distance of the cave to the borehole wall is *L*, the radius of the borehole is *a*, *a* = 10 cm, and the source and receiver are placed in the borehole. The source is located under the cave; the vertical distance from the cave is 2 m. Perfect-matched layer (PML) is an absorbing boundary region. *r* is the radial coordinate and *z* is the axial coordinate of the borehole.

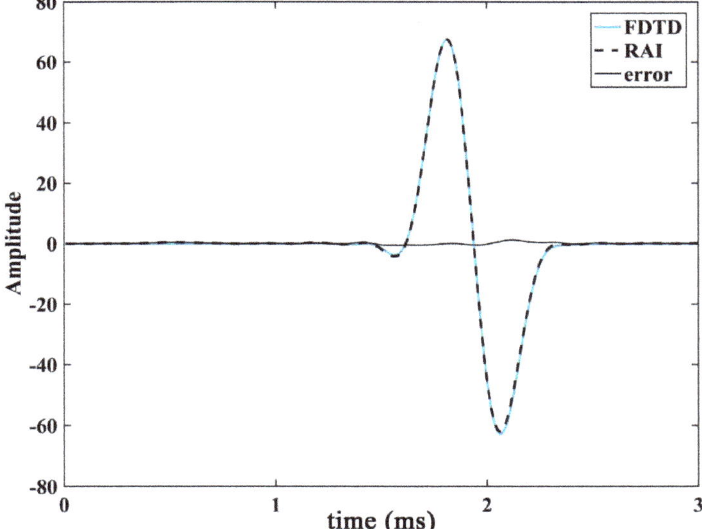

Figure 3. Stoneley (ST) waveforms propagating in a borehole surrounded by a homogeneous formation. The solid blue curve is a waveform obtained by the FD method; the dashed black curve is a waveform obtained by real-axis integration (RAI) method. The solid black curve indicates error.

3. Numeric Simulation Results

3.1. Borehole Acoustic Wave and Wave Field Snapshot of Cave Formation

Figure 4 shows the full-wave waveform of cave formation. In this study, a monopole source is selected. The P and S-wave cannot be excited when the center frequency of the source is 2 kHz. There is only a ST wave in the borehole. The ST wave is unchanged because the cave in the formation cannot directly affect it in Figure 4a. The P, S and pseudo-Rayleigh waves appear in the wave train (Figure 4b–d) with the increase of the source frequency. Scattering waves are formed when the P and S-waves encounter the water-filling cave in the formation. Borehole acoustic waves are also changed because of the superposition with scattered waves [30]. In order to suppress pseudo-Rayleigh wave, the center frequency of source is usually 8–12 kHz.

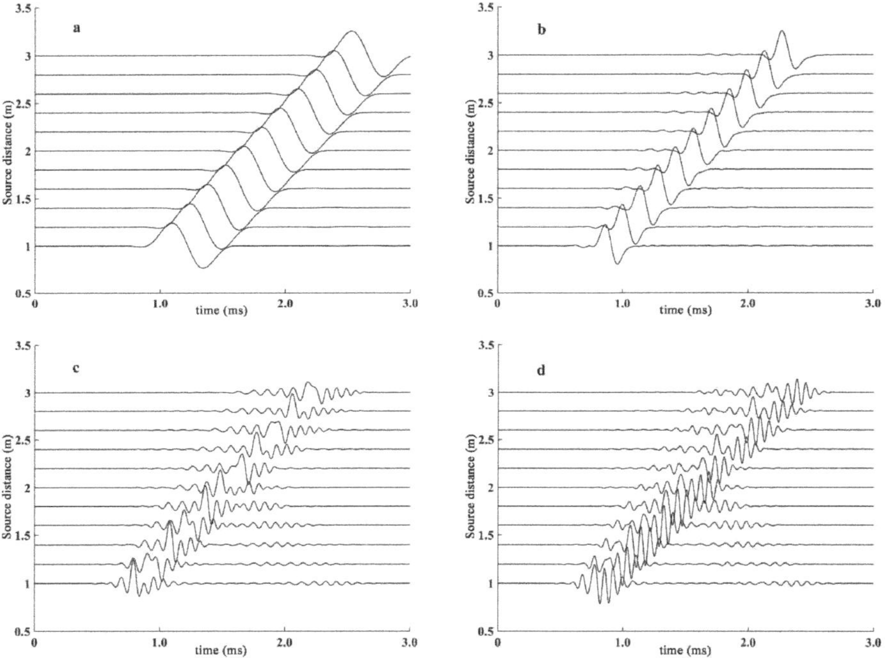

Figure 4. Borehole acoustic wave formed by different excitation frequency. Center frequency of source in (**a**–**d**) are 2 kHz, 5 kHz, 8 kHz and 10 kHz, respectively. Radius of the cave is 2 cm and the distance of the cave to the borehole wall is 2 cm.

The center frequency of monopole source was 10 kHz in this study. Figure 5. shows the wave field of acoustic wave at different times in cave formation. The acoustic wave propagated along the borehole or formation to the PML boundary of the model. There was no artificial boundary reflection, and this proves that the NPML was effective. The P-wave encountered the cave at 0.9 ms. Because the scattered wave formed by P-wave was weak, only the original wave field excited by source can be seen in Figure 5a. The S–P and S–S waves were formed when an S-wave encountered a cave due to energy conversion. The S–P wave can be observed in Figure 5b beside the original wave field formed by the source at 1.3 ms. The scattered-wave field formed by the S–wave can be clearly observed in Figure 5c—including S–P and S–S waves. It should be noted that the amplitude of the S-wave decreased significantly after it passed through the cave in Figure 5c (the color of the S-wave became lighter).

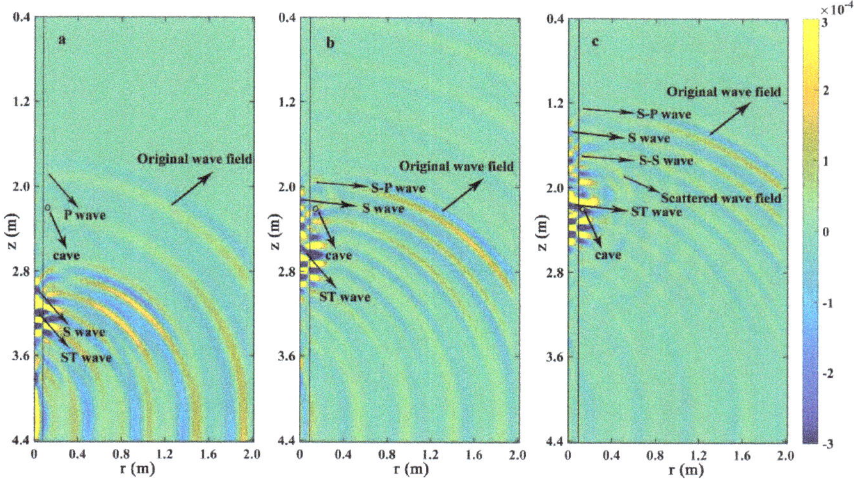

Figure 5. Wave-field snapshot of acoustic waves in a cave formation. (**a–c**) Wave fields at 0.9, 1.3 and 1.6 ms, respectively. Bar scale represents the amplitude of the wave.

3.2. The Effect of the Cave on the Borehole P-Wave

Figure 6a shows the effect of different radii of cave on P-waves.

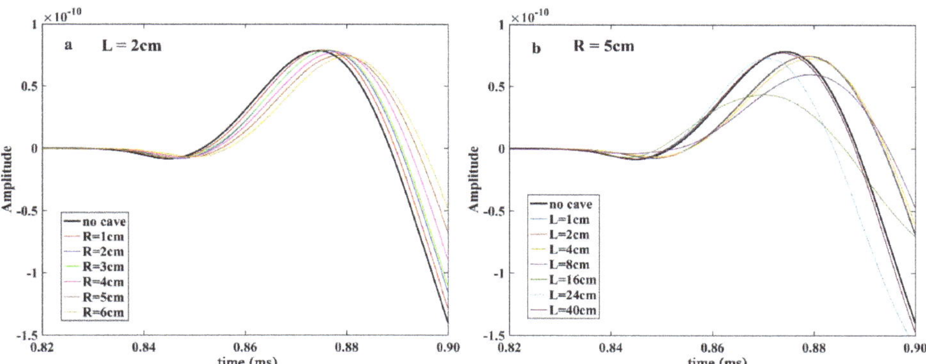

Figure 6. Effect of the cave on P-wave. (**a**) shows the effect of the cave radius on P-wave. (**b**) shows the effect of the cave position on P-wave. R represents the radius of the cave. L represents the distance of the cave to the borehole wall. Center frequency of source is 10 kHz. P-wave recorded by the receiver is 0.4 m above the cave.

1. The amplitude of the P-wave varied slightly with the radius of the cave;
2. The larger the cave radius was, the greater was the delay of the first arrival of the P-wave (the first arrival in our study was defined as the maximum of the first phase). From Figure 7a, the relationship between the first arrival of the P-wave and the radius of the cave was linear;

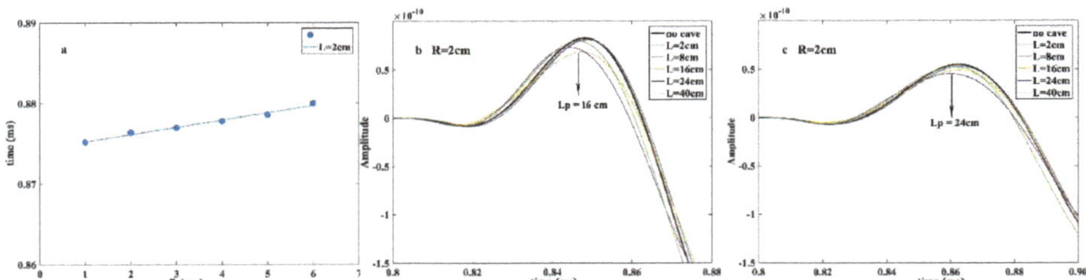

Figure 7. Effect of cave and excitation frequency on the P-wave. (**a**,**b**) Excitation frequency is 10 kHz. (**c**) Excitation frequency is 7 kHz. *R* represents the radius of the cave. *L* represents the distance of the cave to the borehole wall. P-wave recorded by the receiver is 0.4 m above the cave.

3. Figure 6b shows the effect of distance of the cave to the borehole wall on the P-wave;
4. There exists a distance of the cave to the borehole wall defined as L_P. When the distance was equal to L_P, the amplitude of the P-wave generally decreased in observations. Here $L_P = 16$ cm in our model;
5. It was found that L_P was related to the excitation frequency according to the test. $L_P = 24$ cm if the excitation frequency was 7 kHz in Figure 7c and $L_P = 16$ cm if the excitation frequency was 10 kHz in Figure 7b;
6. When the distance was more than 24 cm (the excitation frequency was 10 kHz), it had little effect on the P-wave because of the limitation of investigation depth of the monopole source in our model.

Because the cave was fully filled with water and the wave propagation speed in the fluid was less than that in the formation, the first arrival of the P-wave was delayed. The larger the radius of the cave, the longer was the P-wave propagation time in the fluid, and the later the first arrival time of the P-wave. In this study, a single cave was simulated. There may be multiple caves of different sizes in an actual formation, which will cause more delays on the first arrival of the P-wave [31]. P-wave recorded by the receiver was the superposition of scattered wave and "gliding" P-wave refracted on the borehole wall. The scattered wave formed by the cave changed with different distances to the borehole wall (*L*), so the amplitude of the P-wave may be different when the scattered wave is superimposed with it. In this study, the P-wave amplitude was reduced slightly, and it may increase if the source distance was changed because the scattered wave also varied with different source distance.

3.3. The Effect of the Cave on Borehole S-Wave

Figure 8a shows the effect of the radius of the cave on the S-wave.

1. The attenuation of the S-wave was large (the first arrival in our study was defined as the maximum of the first phase). With the increase of cave radius, the S-wave attenuation became stronger;
2. The energy of the S-wave was more sensitive to cave radius than P-wave;
3. Figure 8b shows the effect of the distance of the cave on S-wave;
4. There exist a distance of cave to the borehole wall defined as L_S. When the distance was equal to L_S, the amplitude of the S-wave decays most by observation. Here $L_S = 16$ cm in our model. L_S was also related to the excitation frequency;
5. When the distance was more than 24 cm (the center frequency of source was 10 kHz), it had little effect on the S-wave in our model. Because of the investigation depth of the monopole source was limited. A distance greater than 24 cm may exceed the maximum investigation depth.
6. S-wave was more sensitive to the change of the distance of the cave than P-wave.

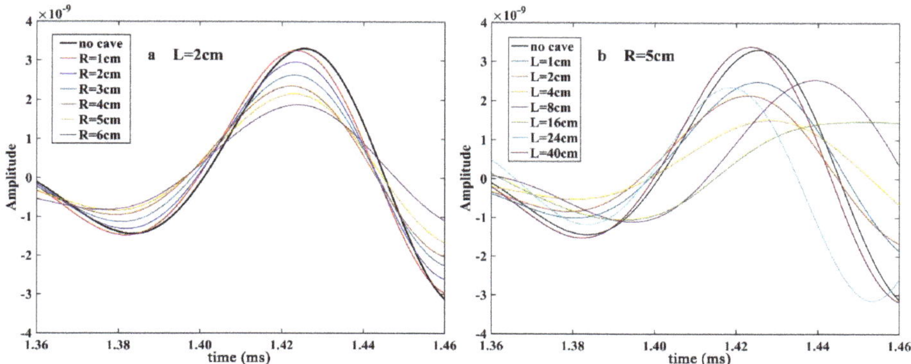

Figure 8. Effect of the radius and position of cave on S-wave. (**a**) shows the effect of the cave radius on S-wave. (**b**) shows the effect of the cave position on S-wave. *R* represents the radius of the cave and *L* represents the distance of the cave to the borehole wall. Center frequency of source is 10 kHz. S-wave recorded by the receiver 0.4 m above the cave.

When the S-wave encounters the water-filled cave, the water inhibits the propagation of the S-wave and a considerable part of the energy of the S-wave is converted into the compressed energy (the energy of S–P wave) of the fluid [31]. The compression energy may be converted into S-wave and other mode waves, and these converted waves are recorded by the receiver finally. Therefore, the S-wave energy recorded by the receiver above the cave will be partially attenuated due to scattering and conversion.

3.4. The Relationship between the Attenuation Coefficient of S-Wave and the Cave

It is found that the existence of a cave can make the S-wave energy decay through simulation and analysis. In this study, the attenuation coefficient A of the S-wave under the action of the cave is defined as

$$A = 1 - \frac{A_2}{A_1} \tag{13}$$

where A_1 and A_2 are the amplitude of the S-wave for a homogeneous formation and a cave formation in the same source distance, respectively. The dotted line is a fitting linear equation in Figure 9a, and there is a positive linear correlation between the attenuation coefficient and the radius of the cave. The slope of linear equation of attenuation coefficient is variable when the distance of cave is different. The large slope indicates that the attenuation coefficient is sensitive to the radius of the cave. When $L = 4$ cm, the attenuation coefficient is the most sensitive to the cave radius in our model.

When the S-wave propagates in the formation and run into the water-filled cave, it will form S–P and S–S waves. The S–P and S–S waves recorded by the receiver near the cave are superposed together, so the attenuation coefficient is smaller as shown in Figure 9b. The S–P and S–S waves are separated gradually with the increase of the source distance. When the S–P leaves the S-wave, it takes its energy away from the S-wave, so the attenuation coefficient increases [31]. The effect of the cave on the S-wave is reduced with the increase of the source distance, so the attenuation coefficient decreases gradually. In our model, the S–S wave was separated completely when the source distance is 2.6 m as can be seen from Figure 9b.

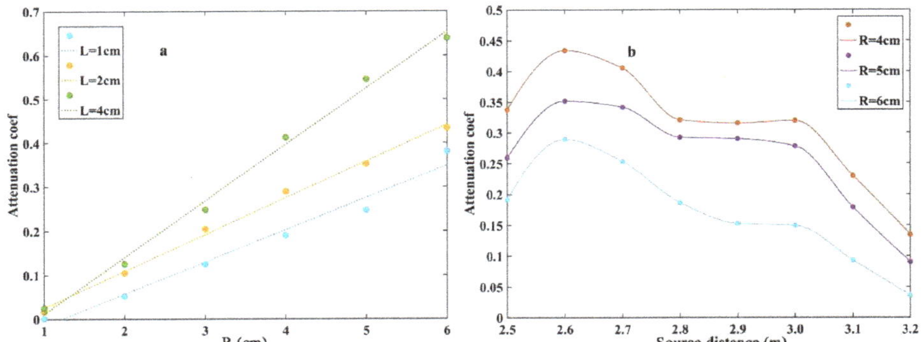

Figure 9. Effect of cave and source distance on attenuation coefficient of the S-wave. (**a**) shows the effect of the cave radius on attenuation coefficient. (**b**) shows the effect of the source distance on attenuation coefficient. R represents the radius of the cave and L represents the distance of the cave to the borehole wall. Center frequency of source is 10 kHz.

3.5. The Effect the of Cave on Stoneley Wave

The P and S-wave which can propagate in the formation are formed when the center frequency of source is high. At this time, the scattered wave propagates to the borehole and affects the ST wave. Figure 10a is the wave train in homogeneous elastic formation. Figure 10b–f represent the wave train in the formation with cave of different radii. It can be seen that when the radius of the cave is small (Figure 10b,c), the attenuation of the ST wave is weak compared with Figure 10a. When the radius of the cave is large (Figure 10d–f), the amplitude of the ST wave has a great attenuation. Therefore, the attenuation of the ST wave not only corresponds to fractures, but also to caves.

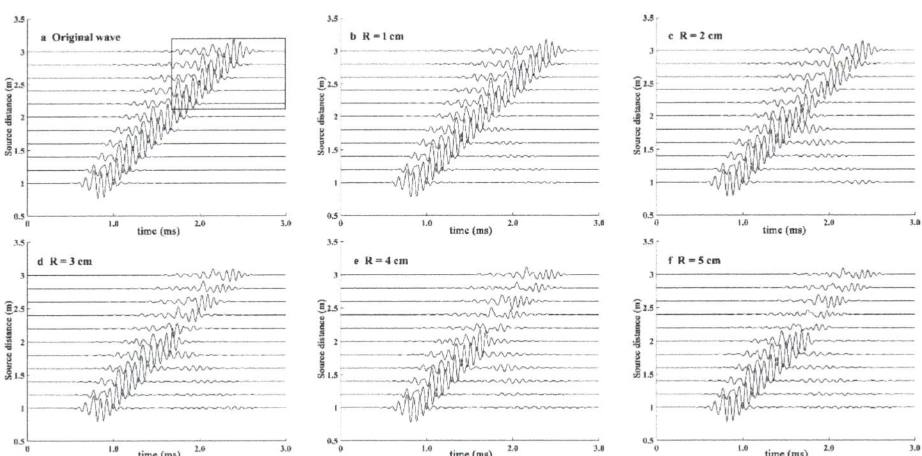

Figure 10. Effect of the radius of the cave on ST wave. (**a**) shows the wave train in homogeneous elastic formation. (**b–f**) represent the wave train in the formation with cave of different radii. R represents the radius of the cave. The distance of the cave to the borehole wall is 2 cm. Center frequency of source is 10 kHz.

Figure 11a shows acoustic waveforms. In order to study the relationship between the ST wave and the cave radius, it is necessary to calculate the energy of the ST wave. The full-wave frequency spectrum is obtained by Fourier transform in Figure 11b; the low-pass filter is used to obtain ST wave. The energy of the ST wave is obtained by summing the square of the corresponding value of each frequency. The frequency of the ST wave is less than 4 kHz according to Figure 11b. The cutoff frequency of the low-pass filter is 4 kHz. The black line in Figure 11a is the low-pass filtered waveform. The S-wave and pseudo-Rayleigh wave have been suppressed to a certain extent. However, due to the complexity of acoustic wave and the limitation of Fourier transform [32], S and pseudo-Rayleigh waves are difficult to be completely suppressed. Figure 11c represents the relationship between ST wave energy and cave radius at different source distances. The energy of the ST wave gradually decreases with the increase of cave radius at different source distances.

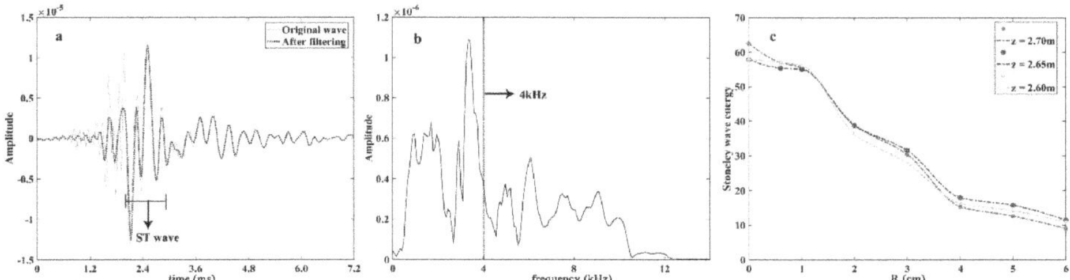

Figure 11. Calculation of ST wave energy and its relationship with cave radius. (**a**) shows the original and the filtered waves. (**b**) shows the frequency spectrum of original wave. (**c**) shows the effect of cave radius on the energy of ST wave. Center frequency of source is 10 kHz. R represents the radius of the cave. z is the axial coordinate.

3.6. The Effect of the Cave on the Scattered Wave

Figure 12a refers to the acoustic wave and the scattered wave. The wave propagation in the borehole of homogeneous formation is regarded as a direct wave, and the wave trains of cave formation in Figure 12a subtract the direct waves to separate the scattered waves. A Fourier transform was used to get the frequency spectrum of the scattered wave in Figure 12b. The peaks with larger amplitude are mainly S–S wave, and the dominant frequency is about 9–12 kHz. The peaks with smaller amplitude are mainly converted P-waves. We added the square of the value of each frequency to get the energy of the scattered wave. Taking the distance and radius of the cave as the abscissa and scattered wave energy as the ordinate, the relationship between the scattered wave energy and the cave was obtained in Figure 13.

The scattered wave recorded at 0.4 m below the cave (the source distance was 1.6 m) was selected. The cave with a diameter less than 0.25 times the wavelength of the S-wave was approximated as a point scatterer; otherwise the cave itself could form multiple-scattering [30]. In our model, a cave with a radius less than 2 cm could be regarded as a point-scatterer and could form multiple-scattering with a radius greater than 3 cm.

Figure 13a shows the relationship between the energy of scattered wave and the position of cave. When the cave can be regarded as a point-scatterer ($R \leq 2$ cm), the energy of the scattered wave has a negative correlation with the distance of the cave. The energy of scattered wave is very small when the distance is more than 24 cm and reaches the maximum when $L = 2$ cm if the cave itself can form multiple-scattering ($R > 3$ cm). The scattered wave energy gradually decreases when $L > 2$ cm ($R > 3$ cm).

Figure 13b shows the relationship between the energy of scattered wave and the radius of the cave. The energy of scattered wave raises with the increase of cave radius if $R \leq 2$ cm. If $R > 3$ cm, the energy of the scattered wave changes sharply when the cave is close to the borehole wall ($L = 2$ cm) and is relatively stable when L > 2 cm with the increase of cave radius.

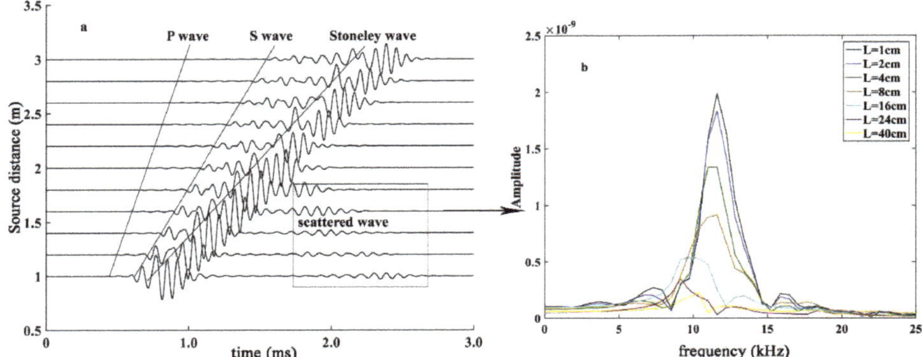

Figure 12. Full-wave acoustic-logging in cave formation. (**a**) shows the scattered wave in the cave formation. (**b**) shows the frequency spectrum of the scattered wave. Radius of cave is 2 cm. *L* represents the distance of the cave. Center frequency of source is 10 kHz.

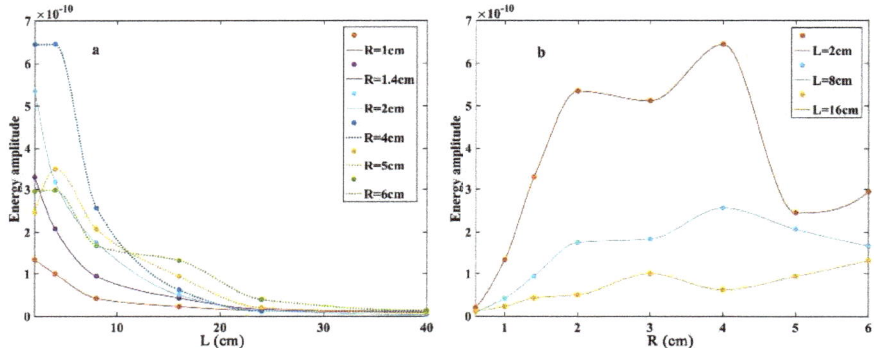

Figure 13. The relationship between the energy of the scattered wave and the cave. (**a**) shows the effect of cave position on the energy of the scattered wave. (**b**) shows the effect of cave radius on the energy of the scattered wave. *R* represents the radius of the cave. *L* represents the distance of the cave. Center frequency of source is 10 kHz.

4. Discussion

The numeric simulation of borehole acoustic waves in cave formations is a theoretical calculation; its ultimate purpose is to provide interpretation and reference for actual data. The results in our study show that borehole full-waves have response characteristics to the caves, which are mainly reflected in the energy and the first arrival of waves.

The porosity-frequency spectrum is calculated from the conductivity curve of electrical-imaging logging [33] and follows the following rules.

(1)   When the primary pores are well developed and the secondary pores are poorly developed, the porosity-distribution spectrum is a narrow single peak. The peaks move to the direction of small porosity;

(2)   When the primary pores are poorly developed and the secondary pores are well developed, the porosity-distribution spectrum is also a single peak. However, the peaks move to the direction of large porosity;

(3) When the primary pores and secondary pores are well developed, but heterogeneous, the porosity-distribution spectrum shows multiple peaks. The peaks move to the direction of large porosity.

Cumulative porosity curves PS3, PS5, PS7 ... PS30 represent pore components with porosity values below 0.03, 0.05, 0.07 ... 0.3 [34]. The porosity of electrical imaging can be obtained by filling the profile with different colors; the contribution of the different sizes of pores to the total porosity can be intuitively analyzed. According to the method in Figure 11b mentioned above, the energy of the ST wave is calculated. The acoustic full-wave was decomposed into several intrinsic mode function (IMF) components by empirical mode decomposition (EMD). P and S-wave were distributed in the highest frequency IMF1 component [35]. The energy of IMF1 component can be approximately regarded as the energy of the S-wave because the amplitude of the P-wave is very small. The slowness of the P-wave is the reciprocal of velocity. The first arrival delay of the P-wave indicates that the velocity is low, then the slowness is large.

Figure 14 shows the logging results of H-well in the igneous formation of Liaohe. Area **a** is basalt. Cores 1–2 show no alteration. The color of electrical-imaging logging is bright, indicating that the conductivity is low. The porosity-distribution spectrum shows a single peak moving forward, and the porosity of electrical imaging shows that small pores are mainly developed in the formation. The energy of the S and ST waves is high and stable. The density is large, and the porosity and slowness of the P-wave are small. All the above logging data indicate that this area is a tight formation. Cores 3–4 show altered basalt in area **b**. Dark spots appear in the electrical-imaging logging static image, which indicates that as conductivity increases, and caves develop. The porosity-distribution spectrum widens and moves in the direction of large porosity. It was found that there are mainly medium and large pores from the electrical imaging porosity. The energy of the S and ST waves was attenuated, and the slowness of the P-wave increased. This is in good agreement with the theoretical results of the cave formation simulated.

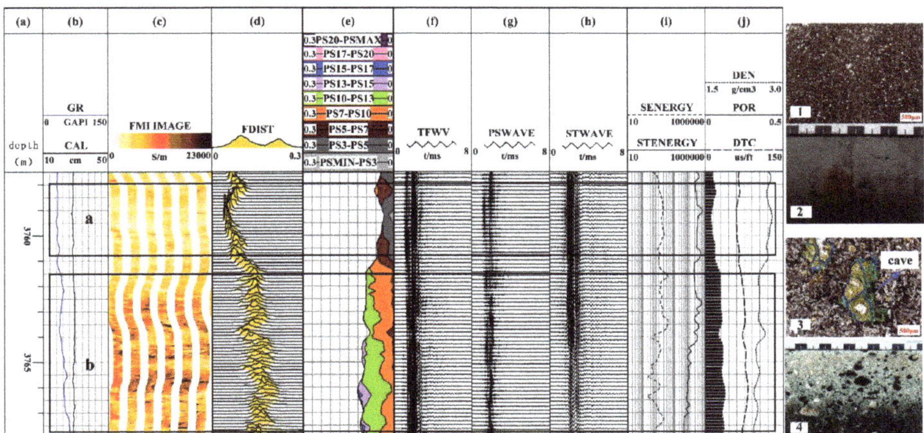

Figure 14. Comprehensive interpretation of logging data in Liaohe igneous formation. (**a**) Depth of the well; (**b**) GR is natural gamma and CAL is caliper; (**c**). FMI IMAGE represents electrical-imaging logging static image; (**d**) FDIST is the porosity-distribution spectrum; (**e**) PSMIN–PSMAX are the porosity of electrical imaging; (**f**–**h**) TFWV, PSWAVE and STWAVE are the waveforms of full-wave, P, S and ST waves, respectively; (**i**) SENERGY represents the energy of the S-wave. STENERGY is the energy of ST wave; (**j**) DEN is the density of the formation. POR is the porosity by conventional logging. DTC is the slowness of the P-wave. Figures 1 and 2 on the right are the core slice and core of the tight formation. Figures 3 and 4 are the core slice and core of the altered formation.

Figure 15 shows the effect of caves with different development degree on borehole acoustic waves. The attenuation of the S and ST waves becomes stronger, and the slowness of the P-wave becomes larger with an increasing number and size of caves.

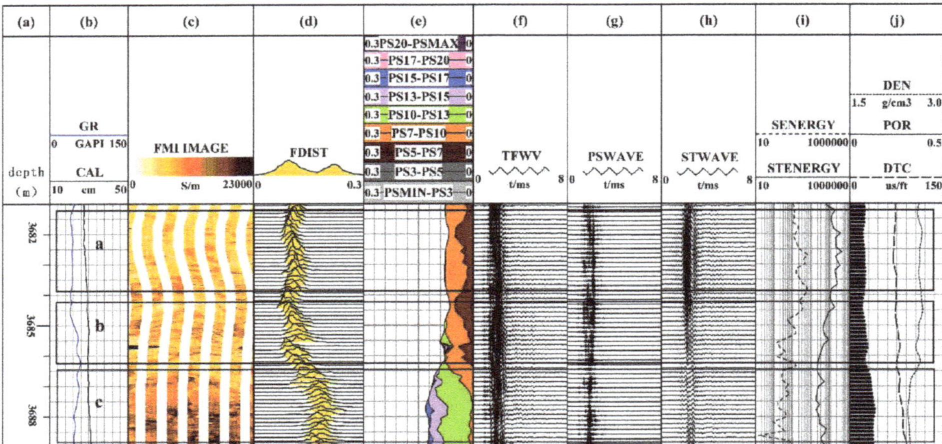

Figure 15. Comprehensive interpretation of logging data of cave formations with different development degrees. (**a**) Depth of the well; (**b**) GR is natural gamma and CAL is caliper; (**c**). FMI IMAGE represents electrical-imaging logging static image; (**d**) FDIST is the porosity-distribution spectrum; (**e**) PSMIN–PSMAX are the porosity of electrical imaging; (**f–h**) TFWV, PSWAVE and STWAVE are the waveforms of full-wave, P, S and ST waves, respectively; (**i**) SENERGY represents the energy of the S-wave. STENERGY is the energy of ST wave; (**j**) DEN is the density of the formation. POR is the porosity by conventional logging. DTC is the slowness of the P-wave.

The theoretical results of this study can be applied to practical data, but there are also some shortcomings.

(1) The model simulated in our study is an ideal simplified model, which has a certain effect in describing the response characteristics of acoustic wave to a single cave. However, the number, shape and filling degree of caves in the actual formation are very complex. The author's future work is to establish a 3D complex geological model and study the response characteristics of acoustic-logging more detailed and systematic;

(2) The actual well-logging data can reflect the development degree of cave, but cannot accurately reflect the details of cave. It is a long way to go that how to use monopole acoustic-logging to detect reservoir caves precisely. More complex numeric simulation and comparison with actual data are needed;

(3) A single GPU parallel-computing program is designed in this study. The memory of a single GPU may not meet the computing needs when the amount of data are too large, so we need to add support for multiple GPUs in the program. It is of great significance to apply CUDA parallel-computing of multiple GPUs to borehole acoustic numeric simulation.

5. Conclusions

The following important conclusions can be drawn:

1. The investigation depth of monopole source was about 24 cm in our model;
2. The cave cannot affect ST wave under the condition of low frequency. The position and size of cave affected the borehole acoustic wave at high frequency;

3. Due to the superposition of scattered waves, the amplitude of the P-wave may increase or decrease. Theoretically, the first arrival of the P-wave can reflect the development degree of cave;

4. The energy of the S-wave is always attenuated due to the propagation mechanism. S-wave is more sensitive to the presence of cave. The energy attenuation of the S-wave can reflect the size of the cave;

5. The attenuation coefficient of the S-wave raises first, and then decreases due to energy conversion with the increase of source distance;

6. For ST waves the amplitude decreases with the increase of cave radius;

7. The excitation frequency may affect the relationship between the distance of the cave and P(S) waves;

8. The energy of the scattered wave generally decreases with an increase of distance to the cave. It is worth noting that the scattered wave formed by large-scale near-borehole cave has no rule with radius and distance of the cave.

Author Contributions: F.X. is the main author of this manuscript and this work was conducted under the advisement of Z.W.; Z.W. is the supervision, project administration. Z.W. reviewed the manuscript and made contributions to its structure. All authors have read and agreed to the published version of the manuscript.

Funding: This research was funded by the National Natural Science Foundation of China, Grant Number: 41874135.

Conflicts of Interest: The authors declare no conflicts of interest.

References

1. Yu, C.M.; Zheng, J.P.; Tang, Y.; Yang, Z.; Qi, X.F. Reservoir Properties and Effect Factors on Volcanic Rocks of Basement beneath Wucaiwan Depression, Junggar Basin. *Earth Sci.-J. China Univ. Geosci.* **2004**, *29*, 303–308.

2. Wen, L.; Li, Y.; Yi, H.Y.; Liu, X.; Zhang, B.J.; Qiu, Y.C.; Zhou, G.; Zhang, X.H. Lithofacies and reservoir characteristics of Permian volcanic rocks in the Sichuan Basin. *Nat. Gas Ind.* **2019**, *39*, 17–27. [CrossRef]

3. Si Ma, L.Q. Study on Comprehensive Evaluation Method and Application of Well Logging for Fracture-Vuggy Carbonate Reservoir. Ph.D. Thesis, Southwest Petroleum University, Chengdu, China, 2005; pp. 11–25.

4. Wang, J.J.; Hu, Y.; Liu, Y.C.; He, P.W.; Lan, X.M.; Wen, W. Dentification and characterization of multi-scale pores, vugs and fractures in carbonate reservoirs: A case study of the Middle Permian Qixia dolomite reservoirs in the Shuangyushi Structure of the northwestern Sichuan Basin. *Nat. Gas Ind.* **2019**, *40*, 48–57.

5. Hei, C. Acoustic Logging Scattering Effect Basic Application Research in Complex Formation. Ph.D. Thesis, China University of Petroleum (East China), Beijing, China, 2016; pp. 1–10.

6. Ou, W.M.; Wang, Z.W. Simulation of Stoneley wave reflection from porous formation in borehole using FDTD method. *Geophys. J. Int.* **2019**, *217*, 2081–2096. [CrossRef]

7. Ou, W.M.; Wang, Z.W.; Ning, Q.Q.; Yu, Y.; Xu, F.H. Numerical simulation of borehole Stoneley wave reflection by a fracture based on variable grid spacing method. *Acta Geophys.* **2019**, *67*, 1119–1129. [CrossRef]

8. Liu, D.; Fan, J.C.; Wu, S.N. Acoustic Wave-Based Method of Locating Tubing Leakage for Offshore Gas Wells. *Energies* **2018**, *11*, 3454. [CrossRef]

9. Burns, D.R.; Willis, M.E.; Toksoz, M.N. Fracture properties from seismic scattering. *Lead. Edge* **2007**, *26*, 1186–1196. [CrossRef]

10. Wu, R.S.; Aki, K. Scattering characteristics of elastic waves by an elastic heterogeneity. *Geophysics* **1985**, *50*, 582–595. [CrossRef]

11. Tang, X.M.; Wei, Z.T. Single-well acoustic reflection imaging using far-field radiation characteristics of a borehole dipole source. *Chin. J. Geophys.* **2012**, *55*, 2798–2807. (In Chinese)

12. Tang, X.M.; Cao, J.J.; Wei, Z.T. Radiation and reception of elastic waves from a dipole source in open and cased boreholes. *J. China Univ. Pet.* **2013**, *37*, 57–64.

13. Tang, X.M.; Cao, J.J.; Wei, Z.T. Shear-wave radiation, reception, and reciprocity of a borehole dipole source: With application to modeling of shear-wave reflection survey. *Geophysics* **2014**, *79*, 43–50. [CrossRef]

14. Hei, C.; Tang, X.M.; Li, Z.; Su, Y.D. Elastic wave scattering and application in the borehole environment. *Sci. Sin.* **2016**, *46*, 118–126.

15. Li, D.; Qiao, W.X.; Che, X.H.; Ju, X.D. Study on the response of dipole reflected acoustic imaging the near-borehole cave. *J. Appl. Acoust.* **2019**, *28*, 767–773.

16. Fan, Z.J.; Liu, J.H.; Zhang, W.F. Log interpretation and evaluation of the Ordovician carbonate rock reservoirs in Tahe oilfield. *Oil Gas Geol.* **2008**, *29*, 61–65.

17. Zhao, J.; Li, Z.J.; Yu, B.; Fan, Z.J. Acoustic Logging Numerical Simulation and Quantitative Evaluation of Cave Filling Degree. *J. Basic Sci. Eng.* **2013**, *21*, 1070–1077.

18. Wang, X.C.; Zhang, J.; Li, J.; Hu, S.; Kong, Q.F. Conventional logging identification of fracture-vug complex types data based on crossplots-decision tree: A case study from the Ordovician in Tahe oilfield, Tarim Basin. *Oil Gas Geol.* **2017**, *38*, 805–812.

19. Jarzyna, J.; Bala, M.; Cichy, A. Elastic Parameters of Rocks from Well Logging in Near Surface Sediments. *Acta Geophys.* **2009**, *58*, 34–48. [CrossRef]

20. Wawrzyniak, K. Application of Time-Frequency Transforms to Processing of Full Waveforms from Acoustic Logs. *Acta Geophys.* **2009**, *58*, 49–82. [CrossRef]

21. Hellwig, O.; Geerits, T.W.; Bohlen, T. Accuracy of 2.5 D seismic FDTD modeling in the presence of small scale borehole structures. In Proceedings of the 73rd EAGE Conference and Exhibition Incorporating SPE EUROPEC, Vienna, Austria, 23–27 May 2011.

22. Wang, H.; Tao, G.; Shang, X.F.; Fang, X.D.; Burns, D.R. Stability of finite difference numerical simulations of acoustic logging-while-drilling with different perfectly matched layer schemes. *Appl. Geophys.* **2013**, *10*, 384–396. [CrossRef]

23. Wawrzyniak, K. Acoustic Full Waveforms as a Bridge between Seismic Data and Laboratory Results in Petrophysical Interpretation. *Acta Geophys.* **2016**, *64*, 2356–2381. [CrossRef]

24. Duane, S.; Mete, Y. *CUDA for Engineers: An Introduction to High-Performance Parallel Computing*; Addison Wesley: Boston, UK, 2016; Chapter 4; pp. 39–43.

25. Max, G.; Ty, M. *Professional CUDA C Programming*; Wrox: Birmingham, UK, 2014; Chapter 2; pp. 18–48.

26. Song, B.; Li, W.; Lian, G.X. EFIT simulation of 2D ultrasonic sound field based on CUDA. *J. Beijing Univ. Aeronaut. Astronaut.* **2019**, *45*, 1322–1328.

27. He, L.L.; Jiang, Y.; Ou, Y.; Jiang, S.S.; Bai, H.T. Revised simplex algorithm for linear programming on GPUs with CUDA. *Multimed. Tools Appl.* **2018**, *77*, 30035–30050. [CrossRef]

28. Wang, T.; Tang, X.M. Finite-difference modeling of elastic wave propagation: A nonsplitting perfectly matched layer approach. *Geophysics* **2003**, *68*, 1749–1755. [CrossRef]

29. Yan, S.G.; Xie, F.L.; Gong, D.; Zhang, C.G.; Zhang, B.X. Borehole acoustic fields in porous formation with tilted thin fracture. *Chin. J. Geophys* **2015**, *58*, 307–317. (In Chinese)

30. Mou, Y.G.; Pei, Z.L. *3D Seismic Numerical Simulation of Complex Medium*; Petroleum Industry Press: Beijing, China, 2005; Chapter 8; pp. 125–136.

31. Paillet, F.L. Acoustic propagation in the vicinity of fractures which intersect a fluid-filled borehole. In Proceedings of the SPWLA 21st Annual Logging Symposium. Society of Petrophysicists and Well-Log Analysts, Lafayette, LA, USA, 8–11 July 1980; pp. DD1–DD33.

32. Wang, Z.W.; Liu, J.H.; Nie, C.Y. Time-frequency analysis of array acoustic logging signal based on Choi-Williams energy distribution. *Prog. Geophys.* **2007**, *22*, 1481–1486.

33. Liu, Z.L.; Wang, Z.W.; Zhou, D.P.; Zhao, S.Q.; Xiang, M. Pore distribution characteristics of the igneous reservoirs in the Eastern Sag of the Liaohe depression. *Open Geosci.* **2017**, *9*, 161–173.

34. Zuo, C.J.; Wang, Z.W.; Xiang, M.; Zhou, D.P.; Liu, Z.L.; Yang, F. The radial pore heterogeneity of volcanic reservoir based on the porosity analysis of micro-electric imaging logging. *Geophys. Prospect. Pet.* **2016**, *55*, 449–454.

35. Xu, F.H.; Wang, Z.W.; Liu, J.H.; Ning, Q.Q.; Yu, Y. Acoustic logging information extraction and fractural volcanic formation characteristics based on empirical mode decomposition. *Geophys. Prospect. Pet.* **2018**, *57*, 936–943.

 *energies*

Article

An Immersed Boundary Method Based Improved Divergence-Free-Condition Compensated Coupled Framework for Solving the Flow–Particle Interactions

Pao-Hsiung Chiu [1], Huei Chu Weng [2,*], Raymond Byrne [3], Yu Zhang Che [4] and Yan-Ting Lin [5,*]

[1] Institute of High Performance Computing, Agency for Science, Technology and Research (A*STAR), Singapore 138632, Singapore; chiuph@ihpc.a-star.edu.sg
[2] Department of Mechanical Engineering, Chung Yuan Christian University, Taoyuan City 320314, Taiwan
[3] Centre for Renewables and Energy, School of Engineering, Dundalk Institute of Technology, Dundalk, A91 K584 County Louth, Ireland; Raymond.Byrne@dkit.ie
[4] Department of Mechanical Engineering, Tokyo Institute of Technology, Tokyo 152-8550, Japan; che.y.aa@m.titech.ac.jp
[5] Institute of Nuclear Energy Research, Atomic Energy Council, Taoyuan City 325207, Taiwan
* Correspondence: hcweng@cycu.edu.tw (H.C.W.); yantinglin@iner.gov.tw (Y.-T.L.); Tel.: +886-3-2654311 (H.C.W.); +886-3-4711400 (ext. 3356) (Y.-T.L.)

Abstract: A flow–particle interaction solver was developed in this study. For the basic flow solver, an improved divergence-free-condition compensated coupled (IDFC^2) framework was employed to predict the velocity and pressure field. In order to model the effect of solid particles, the differentially interpolated direct forcing immersed boundary (DIIB) method was incorporated with the IDFC^2 framework, while the equation of motion was solved to predict the displacement, rotation and velocity of the particle. The hydrodynamic force and torque which appeared in the equations of motion were directly evaluated by fluid velocity and pressure, so as to eliminate the instability problem of the density ratio close to 1. In order to effectively evaluate the drag/lift forces acting on the particle, an interpolated kernel function was introduced. The present results will be compared with the benchmark solutions to validate the present flow–particle interaction solver.

Keywords: immersed boundary method; quasi multi-moment method; incompressible Navier–Stokes equation; dispersion-relation-preserving; flow–structure interaction

Citation: Chiu, P.-H.; Weng, H.C.; Byrne, R.; Che, Y.Z.; Lin, Y.-T. An Immersed Boundary Method Based Improved Divergence-Free-Condition Compensated Coupled Framework for Solving the Flow–Particle Interactions. *Energies* **2021**, *14*, 1675. https://doi.org/10.3390/en14061675

Academic Editors: Robert Castilla, Anton Vernet and Antonio F. Miguel

Received: 23 December 2020
Accepted: 13 March 2021
Published: 17 March 2021

1. Introduction

The offshore wind farm is a wind turbine installed in an offshore area where there is a strong wind field for a long period of time. It can avoid the noise, shading and visual obstruction caused by the wind turbine's operation in the land area. Offshore wind farms can also provide more stable wind energy and low turbulence effects. However, it requires consideration of the effects of wind turbulence on the wind turbine and its supporting structure. Recently, attention has also been paid to the offshore floating wind turbine. How to efficiently simulate the interaction between the floating structure and the water/air interface becomes an important research topic.

The coupling analysis of fluids and solids in practical engineering applications usually requires the consideration of complex geometries. These complex geometries are usually in a stationary or transient motion at high Reynolds number flows, such as offshore stationary wind turbines and marine systems. These fluid–structure problems are typically solved by the traditional body-fitted-grid. In order to generate high-quality mesh, a lot of efforts in terms of manpower are always required to refine the mesh at the high curvature region. The computational cost is then increased, especially when dealing with the moving boundary problems. In addition, the quality of the grid is also very crucial for simulation analysis.

In recent decades, the immersed boundary method has been shown to have great potential for modeling the flow–particle interaction problem. It is achieved by virtue of incorporating introduced momentum forcing terms in the equations of motion to predict the displacement, rotation and velocity of the solid object [1]. The main advantage of employing an immersed boundary method for this kind of problem is that the regular fixed Cartesian grid can still be used for simulating the complex time-varying geometries.

However, how to effectively evaluate the drag/lift of the coupling interface and the particle velocity is still a research topic to be refined. Uhlmann [2] proposed incorporating the regularized delta function approach to model the particulate flows, and the improved methods were then proposed by Kempe and Fröhlich [3] and Breugem [4]. Zhang and Gay [5] also proposed a finite element method based on an immersed boundary for fluid–structure interactions. Alternatively, Luo et al. [6] simulated the fluid–particle interactions by direct-forcing based modified immersed boundary method. To further improve the accuracy of velocity in moving solid, Wang et al. [7] introduced the multi-direct forcing method. Yang and Stern [8] also demonstrated a sharp interface-based immersed boundary method by virtue of the direct-forcing approach for solving particulate flows, while Horng et al. [9] exhibited a prediction-correction based direct-forcing immersed boundary projection method for solving fluid–particle interactions.

With the success of finite volume and finite element frameworks, scientists are also motivated to employ an immersed boundary based fluid–particle solver on the Lattice–Boltzmann (LB) framework. Wu and Shu [10] demonstrated a boundary condition-enforced immersed boundary LB scheme to solve particulate flows, while Hao and Zhu [11] revealed an implicit immersed boundary method on the Lattice–Boltzmann framework for fluid–structure interactions. Wang et al. [12] later evaluated three LB schemes to understand their performance when modeling the particulate flows. Zhou and Fan [13] also presented a LB immersed boundary scheme based on the work of Breugem [4]. Zhang et al. [14] further exhibited the particulate immersed boundary method (PIBM) to speed up the calculation of the particulate flow simulations. Coclite et al. [15] proposed kinematic and dynamic forcing strategies for predicting the transport of inertial capsules, and later extend to model the deformable inertial capsules [16].

Due to the fact that the convergence of the Lattice-Boltzmann framework is based on equilibrium equations, each fluid must have its own balance equation for the convergence of the Lattice–Boltzmann method. Furthermore, the definitions of boundary conditions in the Lattice–Boltzmann method is still developing for fitting the physical properties of interfaces. Moreover, some previous studies still only applied the prescribed motion of solid particle for the fluid–solid interaction simulation.

To alleviate the issue of evaluating drag and lift forces, Chiu and Poh [17] have recently successfully incorporated the improved divergence-free-condition compensated coupled (IDFC^2) framework with the direct forcing immersed boundary (DIIB) method [18] for solving the flows with prescribed-motion time-varying geometries. The spurious force oscillation (SFO) can be efficiently alleviated and the calculation is relatively simple. This motivated us to incorporate the IDFC^2 framework with equation of particle motion for simulating the fluid–particle interactions. The ultimate goal for this study is to develop the solver to model the interaction between the floating wind turbine and water/air interface.

In this study, the framework will be extended for simulating the fluid–particle interactions. To prevent the instability problem when the density ratio is close to 1, the drag and lift forces will be directly evaluated from fluid velocity and pressure. An interpolated kernel function to accurately and effectively evaluate the drag/lift forces will be proposed. Details of calculating the equation of motion as well as near-wall treatment when the solid particle is approaching the wall will also be addressed.

This paper is organized as follows. Section 2 presents the governing equations for fluid flows and solid particles. Section 3 present the DIIB method and the methodology for fluid–particle interactions. Section 4 presents the IDFC^2 based flow–particle interaction framework. Verification studies are presented in Section 5. This is followed by presenting the simulated results to show the applicability for the proposed framework. Finally, some concluding remarks are drawn in Section 6.

2. Governing Equations

2.1. Incompressible Navier–Stokes Equations

The non-dimensional primitive-variable Navier–Stokes equations for incompressible flows can be expressed as the following momentum equations and continuity equation:

$$\frac{\partial \vec{u}}{\partial t} + \nabla \cdot \left(\vec{u}\,\vec{u} \right) = \frac{1}{Re} \nabla^2 \vec{u} - \nabla p + \vec{f} \tag{1}$$

$$\nabla \cdot \vec{u} = 0 \tag{2}$$

where $\vec{u}$ is the velocity vector, p is the pressure field, and Re is the Reynolds number. $\vec{f}$ is external force vectors due to particle motions, and is modeled by the present immersed boundary based flow–particle interaction framework.

2.2. Equations of Motion for Solid Particle

The equations of motion for the rigid body can be written as

$$\rho_S V_s \frac{d\vec{U}}{dt} = (\rho_S - \rho_f) V_s \vec{g} + \vec{F_f} \tag{3}$$

$$\frac{d\vec{X}}{dt} = \vec{U} \tag{4}$$

$$I_S \frac{d\vec{\omega}}{dt} = \vec{\psi_f} \tag{5}$$

$$\frac{d\vec{\Omega}}{dt} = \vec{\omega} \tag{6}$$

where the linear velocity vector $(\vec{V})$ and particle position vector $(\vec{X})$ are due to linear motion, while ω and Ω are the angular velocity and orientation due to the angular motion. ρ_S and ρ_f are the density for the solid particle and fluid, respectively. V_s is the volume of the solid particle, $\vec{g}$ is the gravity, and I_S is the inertial moment of the solid particle. $\vec{F_f}$ and $\vec{\psi_f}$ are the hydrodynamic force and torque acting on the solid particle, respectively.

As stated in [2,3], when we evaluated the $\vec{F_f}$ and $\vec{\psi_f}$ by the summation of forcing terms $\vec{f}$ together with the relation of momentum balance over the corresponding fluid domain [2], it will lead to an instability issue when the density ratio close to 1, or $(\rho_S - \rho_f) \approx 0$. This is due to the fact that the term $(\rho_S - \rho_f)$ will appear in the denominator when solving the equations of motion. In order to avoid this issue, hydrodynamic force and torque are directly evaluated in this study:

$$\vec{F_f} = \rho_f \int \underline{\underline{\sigma}} \cdot n \, dA \tag{7}$$

$$\vec{\psi_f} = \rho_f \int r \times (\underline{\underline{\sigma}} \cdot n) \, dA \tag{8}$$

It is noted that by using the present approach, $(\rho_S - \rho_f)$ will be never appear in the denominator, so that there will be no instability problems even when the density ratio id close to 1. In Section 3.1, the present study will introduce methods to accurately and effectively evaluate the two above equations. Furthermore, the implicit midpoint temporal scheme was employed to solve Equations (3)–(6) in the present study.

3. Differentially Interpolated Direct Forcing Immersed Boundary (DIIB) Method

In this section, the DIIB method is employed to model the solid body. Here, we will briefly describe the fundamental idea of the DIIB method.

Assuming there is a solid object S immersed in the Eulerian grid; in order to model the effect of solid object, the direct forcing term $\vec{f}$ is introduced in Equation (1) to satisfy the velocity condition $(\vec{V})$ of the solid object. With the semi-implicit discretization, Equation (1) can be split into the following two equations:

$$\frac{\vec{u}^* - \vec{u}^n}{\Delta t} = -\nabla \cdot \left(\vec{u}\,\vec{u}\right)^* = \frac{1}{Re}\nabla^2 \vec{u}^* - \nabla p \tag{9}$$

$$\frac{\vec{V} - \vec{u}^*}{\Delta t} = \vec{f} \tag{10}$$

It is noted that in practice, the boundary points of the solid object do not always fall into the Eulerian grid points. The direct forcing method will then evaluate the modified velocity on some target Eulerian grid points, followed by obtaining the corresponding momentum forcing terms based on Equation (10) to model the effect of solid object with velocity condition $\vec{V}$.

For the DIIB method, the velocity (u_S) at Eulerian grid point which is inside and close to the solid boundary, will be modified to satisfy the velocity condition u_{IB} at the solid boundary by virtue of interpolated velocity at fluid domain u_f and Taylor series expansion:

$$u_S = 2u_{IB} - u_f \tag{11}$$

Taking Figure 1 as an example, in general u_F does not always lie on the grid point. The evaluation of u_f is done by "advect" u_f from point Q to point A [18]:

$$\frac{\partial u_f}{\partial \tau} + \left(\vec{n}\cdot\nabla\right)u_f = 0 \tag{12}$$

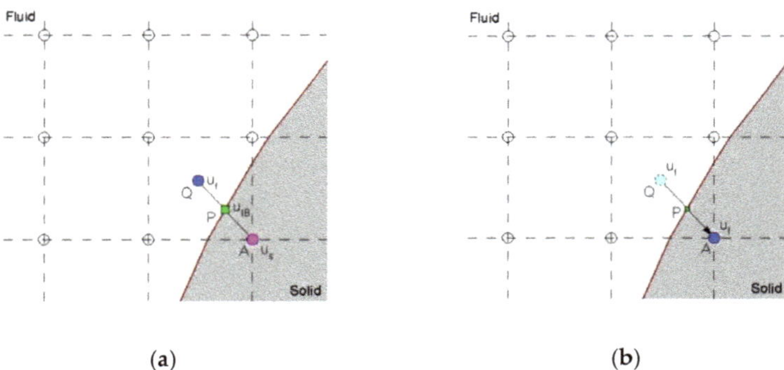

(a)   (b)

Figure 1. Schematic of the direct forcing immersed boundary (DIIB) method: (a) evaluation of u_S; and (b) evaluation of u_f.

In the above, $\vec{n}$ is the unit normal vector of the solid object. In practice, Equation (12) is solved by first order upwind with single artificial time step $\Delta\tau = \overline{APQ}/\left|\vec{n}\right| = 2\overline{AP}/\left|\vec{n}\right|$. The reader can refer to [18] for details.

To alleviate the spurious force oscillations (SFOs) and to obtain smooth pressure, the velocity modification method [17] was employed in this study. The idea behind this method is to modify the velocity inside the velocity to the solid object moving velocity when solving the pressure correction equation. It has been shown in [17] that this method successfully suppresses the SFO for the problems with prescribed motions and will then be implemented for dealing with the problem for which the motions are obtained by solving equations of motion.

3.1. Evaluation of Drag and Lift Forces

When employing the direct forcing immersed boundary method, it is suggested to evaluate forces $\underline{F}$ by the following equation for calculating drag and lift forces:

$$\underline{F} = \int \underline{\underline{\sigma}} \cdot n \, dA \tag{13}$$

where $\underline{\underline{\sigma}}$ is the summation of shear and pressure stress, n is unit normal vector and A is surface area. In this study, a four-point piecewise discrete delta function based interpolated kernel is employed [17]:

$$F(X,Y) = \sum F(x,y)\phi'\left(\frac{x-X}{h}\right)\phi'\left(\frac{y-Y}{h}\right) \tag{14}$$

$$\phi'\left(\frac{x-X}{h}\right) = \phi\left(\frac{x-X}{h}\right) \Big/ \sum \phi\left(\frac{x-X}{h}\right) \tag{15}$$

$$\phi'\left(\frac{y-Y}{h}\right) = \phi\left(\frac{y-Y}{h}\right) \Big/ \sum \phi\left(\frac{y-Y}{h}\right) \tag{16}$$

$$\phi(r) = \begin{cases} \frac{1}{8}\left(3 - 2|r| + \sqrt{1+4|r|+4r^2}\right), & |r| \leq 1 \\ \frac{1}{8}\left(5 - 2|r| - \sqrt{-7+12|r|-4r^2}\right), & 1 \leq |r| \leq 2 \\ 0, & |r| \geq 2 \end{cases} \tag{17}$$

when the solid particle is approaching the wall, the interpolated kernel will be simplified to the following one-point piecewise discrete delta function [19]:

$$\phi(r) \begin{cases} 1 - |r|, & |r| \leq 1 \\ 0, & |r| \geq 1 \end{cases} \tag{18}$$

4. Improved Divergence-Free-Condition Compensated Coupled (IDFC^2) Framework

The idea behind the IDFC^2 framework is to discretize momentum equations by virtue of cell-center and cell-face velocity [17]. With the present approach, we can not only obtain the accurate velocity vector (u, v), but also the fully coupled pressure field p. The following will first introduce the original IDFC framework, then briefly describe how to mitigate to IDFC^2 framework.

4.1. Derivation of IDFC Framework

With the 2D grid structure as shown in Figure 2, the derivation of the IDFC framework starts from the semi-discrete momentum equation. Without loss of generality, we take velocity u at the cell-center P (u_p) as an example:

$$\frac{u_P^* - u_P^n}{\Delta t} + \left(\frac{\partial uu}{\partial x}\right)_P + \left(\frac{\partial vu}{\partial y}\right)_P = \frac{1}{Re}\left(\frac{\partial^2 u}{\partial x^2} + \frac{\partial^2 u}{\partial y^2}\right)_P - \left(\frac{\partial p^n}{\partial x}\right)_P \tag{19}$$

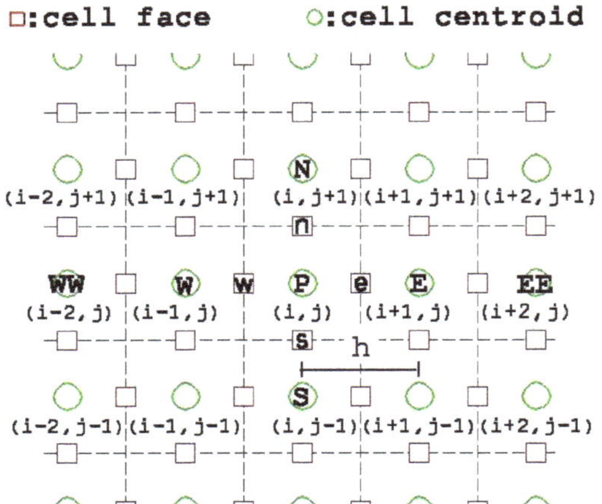

Figure 2. Schematic of the present framework.

The idea for the IDFC method is that discretizing the momentum equation for the cell-face u_e should have the same form as we discretize the momentum equation at the cell-center:

$$\frac{u_e^* - u_e^n}{\Delta t} + \left(\frac{\partial uu}{\partial x}\right)_e + \left(\frac{\partial vu}{\partial y}\right)_e = \frac{1}{Re}\left(\frac{\partial^2 u}{\partial x^2} + \frac{\partial^2 u}{\partial y^2}\right)_e - \left(\frac{\partial p^n}{\partial x}\right)_e \tag{20}$$

However, obtaining the velocity derivatives on the cell-face is not as easy as on the cell-center. The linear-averaged approximation is then employed to evaluate the cell-face velocity derivatives on the cell-face e [20]:

$$\left(\frac{\partial uu}{\partial x}\right)_e \approx \left(\frac{\partial \widehat{uu}}{\partial x}\right)_e = \frac{1}{2}\left(\left(\frac{\partial uu}{\partial x}\right)_E + \left(\frac{\partial uu}{\partial x}\right)_P\right) \tag{21}$$

$$\left(\frac{\partial vu}{\partial y}\right)_e \approx \left(\frac{\partial \widehat{vu}}{\partial y}\right)_e = \frac{1}{2}\left(\left(\frac{\partial vu}{\partial y}\right)_E + \left(\frac{\partial vu}{\partial y}\right)_P\right) \tag{22}$$

$$\frac{1}{Re}\left(\frac{\partial^2 u}{\partial x^2} + \frac{\partial^2 u}{\partial y^2}\right)_e \approx \frac{1}{Re}\left(\frac{\partial^2 u}{\partial x^2} + \frac{\partial^2 u}{\partial y^2}\right)_e = \frac{1}{2}\left(\frac{1}{Re}\left(\frac{\partial^2 u}{\partial x^2} + \frac{\partial^2 u}{\partial y^2}\right)_E + \frac{1}{Re}\left(\frac{\partial^2 u}{\partial x^2} + \frac{\partial^2 u}{\partial y^2}\right)_P\right) \tag{23}$$

By using Equations (21)–(23) into Equation (20), u_e^* can then be obtained. u_w^*, v_n^*, v_s^* can also be obtained in a similar way. In order to fulfill the divergence-free condition, the Poisson equation for pressure correction is then solved with cell-face velocity:

$$\left(\frac{\partial^2 p'}{\partial x^2} + \frac{\partial^2 p'}{\partial y^2}\right)_P = \frac{1}{\Delta t}\left(\frac{u_e^* - u_w^*}{\Delta x} - \frac{v_n^* - v_s^*}{\Delta y}\right) \tag{24}$$

It is followed by updating the cell-face velocity by virtue of the predicted pressure correction:

$$u_e^{n+1} = u_e^* - \Delta t\left(\frac{\partial p'}{\partial x}\right)_e \tag{25}$$

$$u_w^{n+1} = u_w^* - \Delta t\left(\frac{\partial p'}{\partial x}\right)_w \tag{26}$$

$$v_n^{n+1} = v_n^* - \Delta t\left(\frac{\partial p'}{\partial y}\right)_n \tag{27}$$

$$v_s^{n+1} = v_s^* - \Delta t\left(\frac{\partial p'}{\partial y}\right)_s \tag{28}$$

while the pressure is updated as $p^{n+1} = p^n + p'$. Finally, the cell-center velocity is updated by again employing the linear-averaged approximation for the cell-center derivatives:

$$\frac{u_P^* - u_P^n}{\Delta t} + \left(\frac{\partial u\hat{u}}{\partial x}\right)_P + \left(\frac{\partial v\hat{u}}{\partial y}\right)_P = \frac{1}{Re}\left(\frac{\partial^2 u}{\partial x^2} + \frac{\partial^2 u}{\partial y^2}\right)_P - \left(\frac{\partial p^n}{\partial x}\right)_P \tag{29}$$

where:

$$\left(\frac{\partial u\hat{u}}{\partial x}\right)_P = \frac{1}{2}\left(\left(\frac{\partial u\hat{u}}{\partial x}\right)_e + \left(\frac{\partial u\hat{u}}{\partial x}\right)_w\right) \tag{30}$$

$$\left(\frac{\partial v\hat{u}}{\partial y}\right)_P = \frac{1}{2}\left(\left(\frac{\partial v\hat{u}}{\partial y}\right)_e + \left(\frac{\partial v\hat{u}}{\partial y}\right)_w\right) \tag{31}$$

$$\frac{1}{Re}\left(\frac{\partial^2 u}{\partial x^2} + \frac{\partial^2 u}{\partial y^2}\right)_P = \frac{1}{2}\left(\frac{1}{Re}\left(\frac{\partial^2 u}{\partial x^2} + \frac{\partial^2 u}{\partial y^2}\right)_e + \frac{1}{Re}\left(\frac{\partial^2 u}{\partial x^2} + \frac{\partial^2 u}{\partial y^2}\right)_w\right) \tag{32}$$

It is noted that by virtue of Equations (30)–(32), the introduced IDFC forcing terms can efficiently stabilize the calculations [20].

4.2. Derivation of IDFC^2 Framework

It is known from [17] that the IDFC method may lead to non-physical velocity oscillations due to the fact that the cell-face and cell-center velocity is not strongly coupled when evaluating the velocity derivatives. To resolve this issue, we modified Equation (23) to include the contribution of cell-face velocity:

$$\frac{1}{Re}\left(\widetilde{\frac{\partial^2 u}{\partial x^2} + \frac{\partial^2 u}{\partial y^2}}\right)_e = \frac{1}{Re}\left(C\left(\frac{\partial^2 u}{\partial x^2} + \frac{\partial^2 u}{\partial y^2}\right)_e + (1-C)\left(\frac{\partial^2 u}{\partial x^2} + \frac{\partial^2 u}{\partial y^2}\right)_e\right) \tag{33}$$

In the above, C is the newly introduced parameter which is utilized as coupling cell and face velocity. In this study, C was chosen as $1/2$ to ensure the strong coupling. The approximation equation for cell-face velocity u_e^* is then rewritten as

$$\frac{u_e^* - u_e^n}{\Delta t} + \left(\frac{\partial u\hat{u}}{\partial x}\right)_e + \left(\frac{\partial v\hat{u}}{\partial y}\right)_e = \frac{1}{Re}\left(\widetilde{\frac{\partial^2 u}{\partial x^2} + \frac{\partial^2 u}{\partial y^2}}\right)_e - \left(\frac{\partial p^n}{\partial x}\right)_e \tag{34}$$

The corresponding cell-center velocity is expressed as

$$\frac{u_P^* - u_P^n}{\Delta t} + \left(\frac{\partial u\hat{u}}{\partial x}\right)_P + \left(\frac{\partial v\hat{u}}{\partial y}\right)_P = \frac{1}{Re}\left(\widetilde{\frac{\partial^2 u}{\partial x^2} + \frac{\partial^2 u}{\partial y^2}}\right)_P - \left(\frac{\partial p^n}{\partial x}\right)_P \tag{35}$$

$$\frac{1}{Re}\left(\widetilde{\frac{\partial^2 u}{\partial x^2} + \frac{\partial^2 u}{\partial y^2}}\right)_P = \frac{1}{2}\left(\frac{1}{Re}\left(\widetilde{\frac{\partial^2 u}{\partial x^2} + \frac{\partial^2 u}{\partial y^2}}\right)_e + \frac{1}{Re}\left(\widetilde{\frac{\partial^2 u}{\partial x^2} + \frac{\partial^2 u}{\partial y^2}}\right)_w\right) \tag{36}$$

The overall algorithm for the present DIIB–IDFC^2 framework is summarized as follows. The linear and angular velocity, position and rotation of the solid objects are first solved by the equations of motion, where the hydrodynamic force and torque are obtained from the fluid velocity and pressure. To compute the summation of the hydrodynamic force and torque, the solid surface is first divided by a collection of Lagrangian points, and then interpolates the fluid velocity and pressure on each Lagrangian point. Finally, the linear and angular velocity of solid objects are then employed as the boundary condition for the solid–fluid interface. For the sake of completeness, we also plotted the algorithm in Figure 3. In this study, a finite volume-based, second order accurate dispersion-relation preserving upwinding scheme [20] was utilized to discretize the convection terms, while the central difference was employed for other derivative terms. The second order semi-implicit

Gear method was employed as the temporal scheme. For more details of the numerical scheme, people can refer to [20].

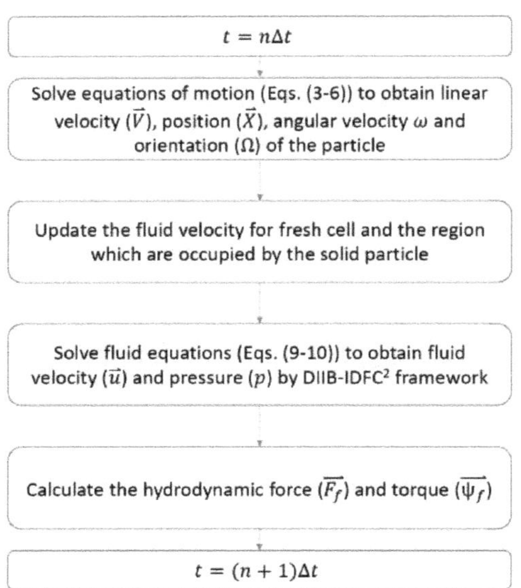

Figure 3. Overall algorithm for the present framework.

5. Results

5.1. Taylor-Couette Flow

The 2-D Taylor–Couette flow problem has been chosen as the first validation problem. For the present problem, the radius for outer circular (R_O) and inner circular (R_I) are 0.4 and 0.2. The non-slip boundary condition is employed for the outer circular, and a constant rotating angular velocity $\omega_{TC} = 3$ is set as a boundary condition for the inner circular. The exact solution for this kind of flow setting can be derived as

$$u(x,y) = -K\left(\frac{R_O^2}{r^2} - 1\right)y \tag{37}$$

$$v(x,y) = K\left(\frac{R_O^2}{r^2} - 1\right)x \tag{38}$$

$$p(x,y) = K^2\left(\frac{r^2}{2} - \frac{R_O^2}{2r^2} - R_2^2 \log\left(r^2\right)\right) \tag{39}$$

$$K = \frac{\omega_{TC} R_I^2}{R_O^2 - R_I^2} \tag{40}$$

$$r = \sqrt{x^2 - y^2} \tag{41}$$

For the present study, four different mesh sizes, namely $\Delta x = \Delta y = \Delta h = 1/10$, $1/20$, $1/40$ and $1/80$ with $Re = 500$ are conducted for the validation study in a unit square domain. The inner and outer cylinder are modeled with DIIB method, as shown in Figure 4.

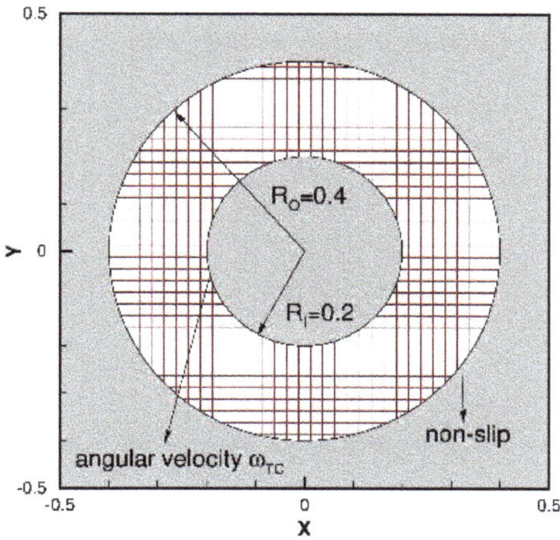

Figure 4. Schematic of the Taylor–Couette flow problem.

From Table 1 and Figure 5, this shows that the rate of convergence for u, v and p are 2.033, 2.033 and 2.044, which match very well with the proposed accuracy order. The validation of the present solver was then confirmed.

Table 1. The error L-2 norms for the Taylor–Couette flow problem.

	u	v	p
$\Delta h = 1/10$	3.380×10^{-2}	3.380×10^{-2}	2.053×10^{-2}
$\Delta h = 1/20$	2.214×10^{-2}	2.214×10^{-2}	9.046×10^{-3}
$\Delta h = 1/40$	1.911×10^{-3}	1.911×10^{-3}	9.115×10^{-4}
$\Delta h = 1/80$	4.928×10^{-4}	4.928×10^{-4}	2.923×10^{-4}

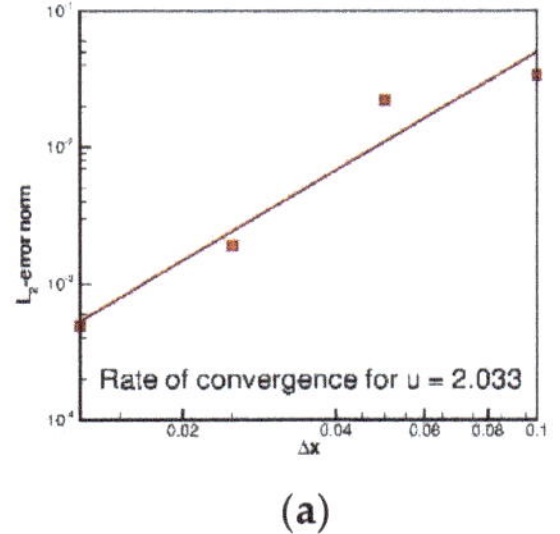

(**a**)

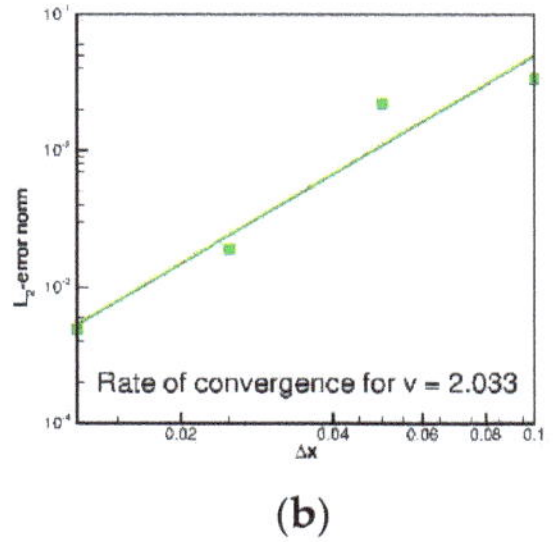

(**b**)

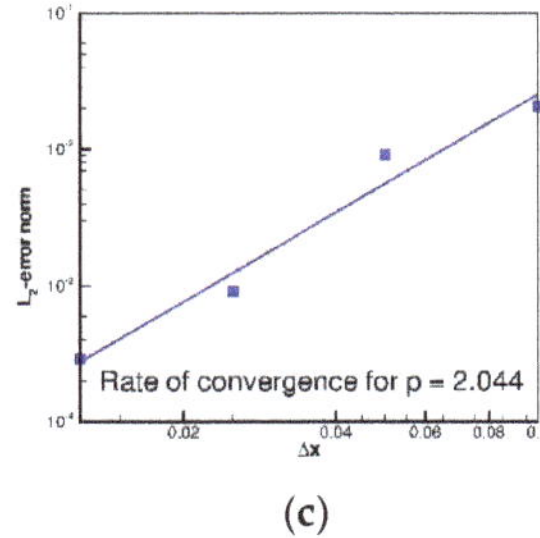

(**c**)

Figure 5. The rate of convergence plot for the Taylor–Couette flow problem. (**a**) u; (**b**) v; and (**c**) p.

5.2. Lid-Driven Semi-Circular Cavity Flow

To further understand the performance of the present framework, the lid-driven semi-circular cavity flow problem was then investigated. In a 1×0.5 rectangular domain, there is a semi-circular cavity with a radius of 0.5 modeled by immersed boundary method, and the lid is driven with velocity $u_{lid} = 1$, as shown in Figure 6. Comparisons of cutting-line velocity at $x = 0.5$ and $y = 0.25$ with the benchmarking solutions of Glowinski et al. [21] and Ding et al. [22] were made with $Re = 1000$ and 100×50 mesh. This shows excellent agreement with Figure 7. To further show the accuracy of the present IDFC^2 framework, we also plotted the numerical results which was obtained by the fifth order upwinding scheme-based DFC framework [23]. As can be seen in Figure 7, the present results match better than the reference results. For the sake of completeness, the predicted contours of pressure and vorticity are also plotted in Figure 8 to show the applicability of the present framework.

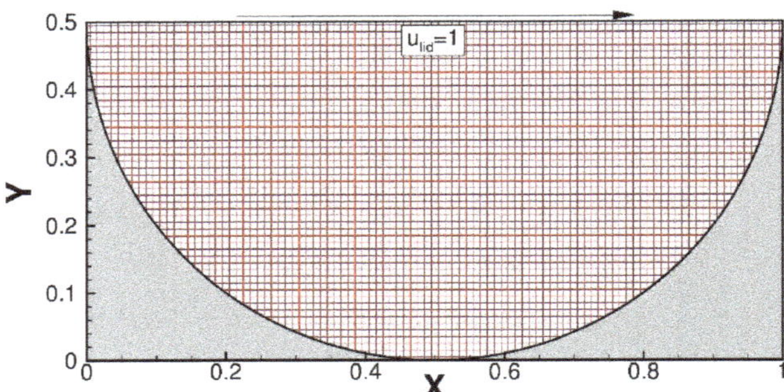

Figure 6. Schematic of the lid-driven semi-circular cavity flow problem.

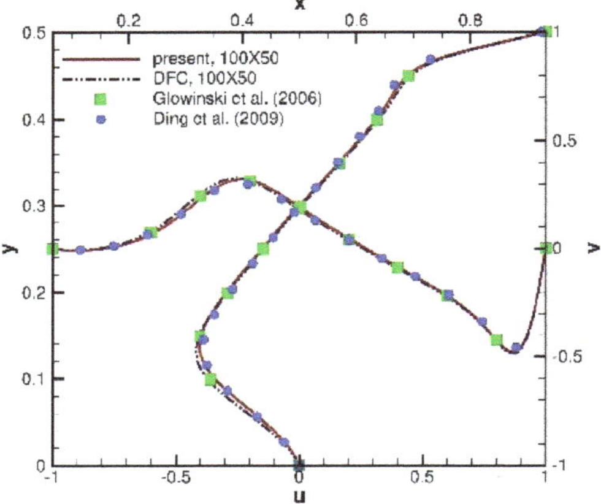

Figure 7. Comparisons of the velocity with the referenced solutions at the cutting lines for the lid-driven semi-circular cavity flow problem.

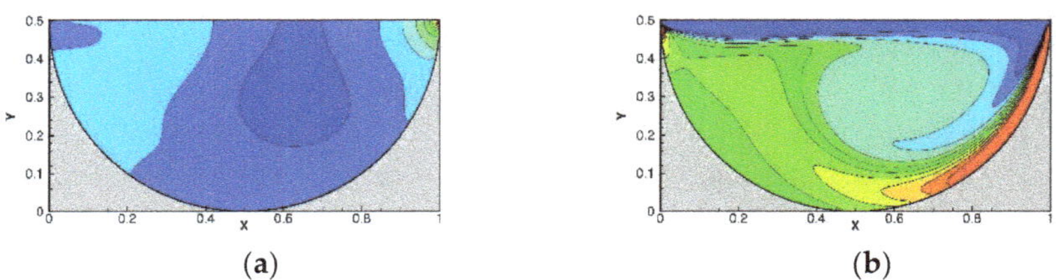

(**a**) (**b**)

Figure 8. The contour plots for the lid-driven semi-circular cavity flow problem: (**a**) pressure; and (**b**) vorticity.

5.3. Flow Past Circular Cylinders in Tandem

In this subsection, the flow past circular cylinders in tandem problem will be simulated to show that the present framework can produce accurate solutions for time transient flows. For this problem the computational domain is set as a 45×20 rectangular domain at the inlet with velocity $u_\infty = 1$ from the left boundary, while the other three boundaries are set as convective boundary conditions. Inside the computational domain, there are two circular cylinders with diameter $d = 1$ in tandem. The centroid of the left one is located at (9.5, 0), while the right one is located at (9.5 + D, 0), where D is the distance between two circular centroids. In this study, two scenarios of D = 4 and D = 5 with $Re = 200$ and mesh size $\Delta x = \Delta y = d/20$ were investigated. The details of the present settings are plotted in Figure 9. The simulated vorticity and pressure contours at t = 400 are plotted in Figures 10 and 11 for D = 4 and D = 5, respectively. As shown in Figures 10 and 11, the present framework can correctly produce the vortex shedding behavior, while the pressure can remain smooth. The predicted results were then used to evaluate the drag coefficient (C_D), lift coefficient (C_L) and Strouhal number (St), and compare with the referenced solutions ([24–27]). From Table 2, good agreements can be observed between predicted results and benchmarking solutions. The drag and lift coefficient are also plotted in Figures 12 and 13 with respect to time. It is confirmed that the present framework for transient flows is applicable and accurate.

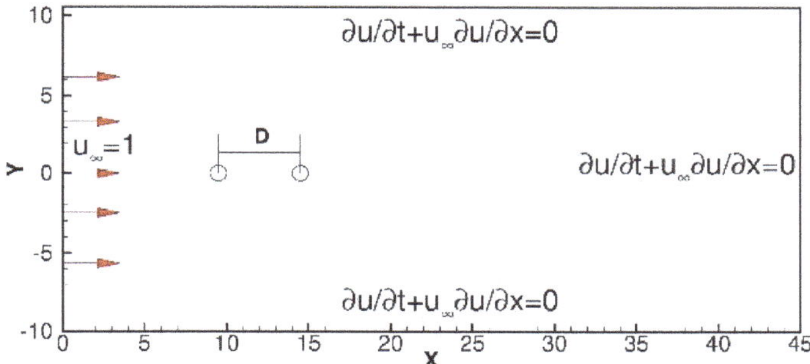

Figure 9. Schematic of the flow past circular cylinders in tandem problem.

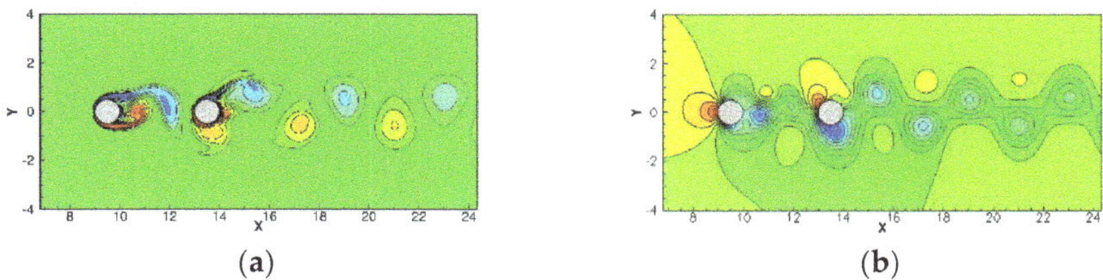

(a) (b)

Figure 10. The instantaneous contour plots for the flow past circular cylinders in tandem problem, D = 4: (**a**) vorticity; and (**b**) pressure.

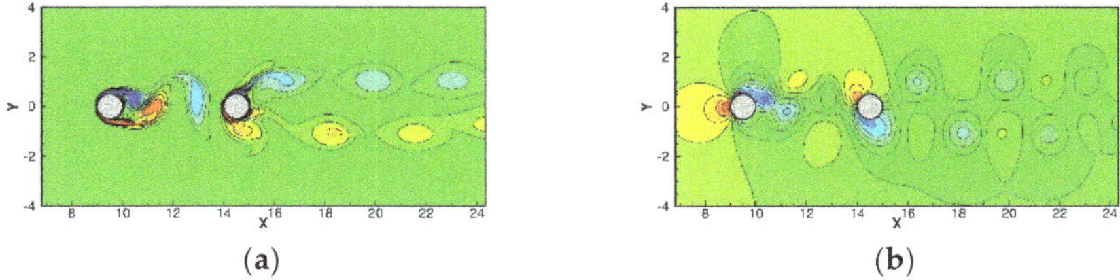

(a) (b)

Figure 11. The instantaneous contour plots for the flow past circular cylinders in tandem problem, D = 5: (**a**) vorticity; (**b**) pressure.

Table 2. Comparisons of C_D, C_L and St for the flow past circular cylinders in tandem problem with $Re = 200$. Note that C_D1 is the left circular cylinder, while C_D2 is the right circular cylinder.

	C_D1	C_D2	C_L1	C_L2	St
D = 4					
Meneghini et al., 2001	1.18	0.38	—	—	0.174
Mahir and Altac, 2008	1.34	0.558	0.805	1.99	0.181
Dehkordi et al., 2011	1.16	0.52	—	—	0.179
Slaouti and Stansby, 1992	1.11	0.88	0.7	1.8	0.190
Present study	1.293	0.568	0.783	1.853	0.183
D = 5					
Mahir and Altac, 2008	1.327	0.455	0.731	1.569	0.186
Slaouti and Stansby, 1992	0.97	0.7	0.55	1.6	0.180
Present study	1.277	0.418	0.702	1.482	0.185

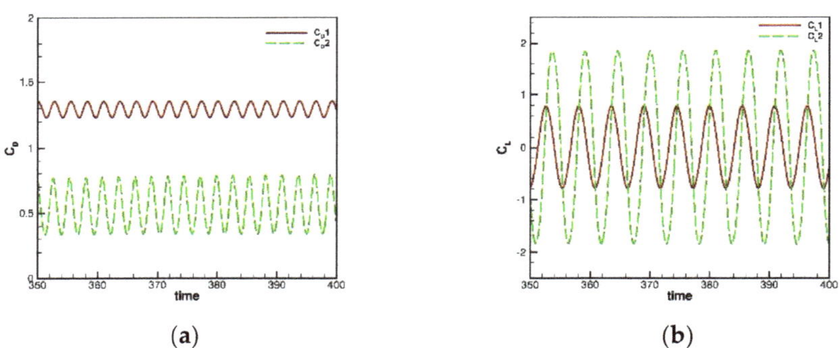

(a)

(b)

Figure 12. The time history plots for the flow past circular cylinders in tandem problem, D = 4. Note that C_D1 is the left circular cylinder, while C_D2 is the right circular cylinder: (**a**) C_D; and (**b**) C_L.

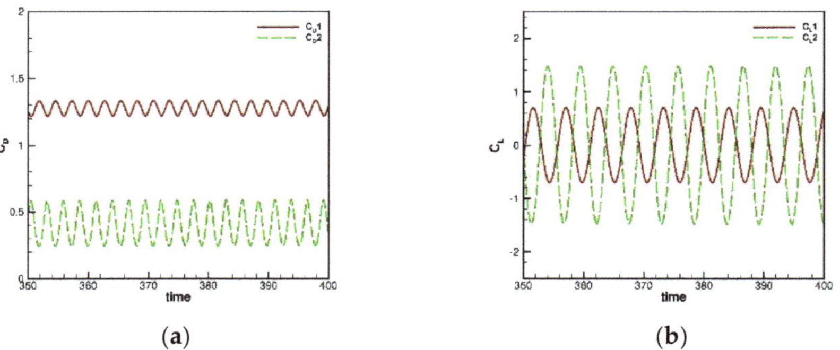

(a)

(b)

Figure 13. The time history plots for the flow past circular cylinders in tandem problem, D = 5. Note that C_D1 is the left circular cylinder, while C_D2 is the right circular cylinder: (**a**) C_D; and (**b**) C_L.

5.4. Two Circular Cylinders Moving towards Each Other in Quiescent Flow

In the computational domain ($x = -8 \sim 24, y = -8 \sim 8$), there are two circular cylinders with a unit diameter (d = 1), the lower initially located at $(x_l, y_l = 0, 0)$, while the upper one located at $(x_u, y_u = 16, 1.5)$, and then moving toward each other along the x direction with the prescribed motions until $t = 32$:

$$x_l(t) = \begin{cases} \frac{4}{\pi} \sin\left(\frac{\pi t}{4}\right), & 0 \le t \le 16 \\ t - 16, & 0 \le t \le 32 \end{cases} \tag{42}$$

$$x_u(t) = \begin{cases} 16 - \frac{4}{\pi} \sin\left(\frac{\pi t}{4}\right), & 0 \le t \le 16 \\ 32 - t, & 0 \le t \le 32 \end{cases} \tag{43}$$

Based on the above prescribed motions, two cylinders will move periodically in the x axis until $t = 16$, then start to move towards each other. In this study, a no-slip boundary condition is applied to all boundaries, and the mesh size and time step size are chosen as $\Delta x = \Delta y = d/40$ and $\Delta t = 10^{-2}$. From Figures 14 and 15, it shows that the present framework can produce smooth pressure and vorticity. To validate the accuracy of the predicted results, we plotted Figure 16 to compare the time-history drag (C_D) and lift (C_L) coefficients of the upper cylinder with the referenced solutions [9,28] which shows excellent agreements. For the sake of completeness, we also plotted Figure 17 to show the effect of velocity modification method. This clearly shows that with the velocity modification method, SFO is well suppressed.

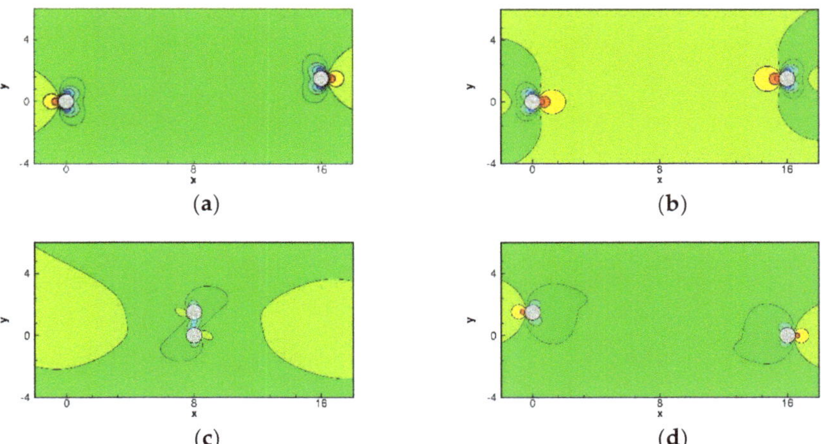

Figure 14. The predicted pressure contour plots for the problem of two cylinders moving towards each other: (a) $t = 4$; (b) $t = 16$; (c) $t = 24$; and (d) $t = 32$.

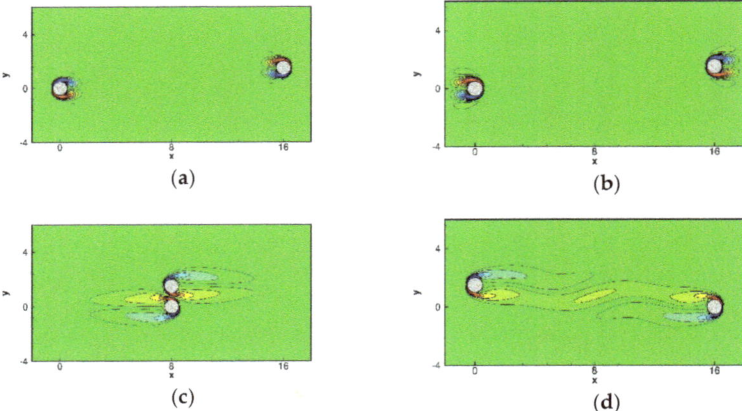

Figure 15. The predicted vorticity contour plots for the problem of two cylinders moving towards each other problem. (**a**) $t = 4$; (**b**) $t = 16$; (**c**) $t = 24$; and (**d**) $t = 32$.

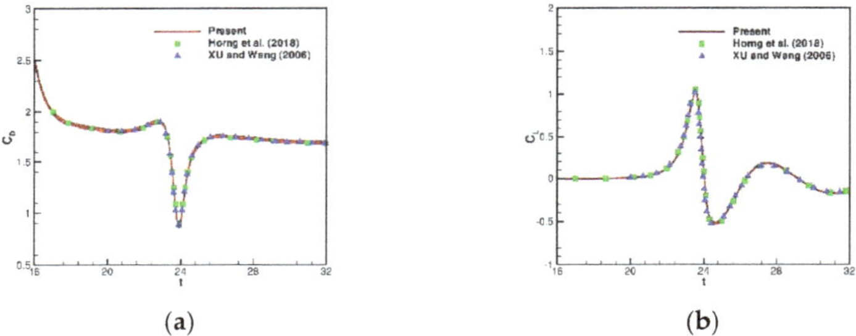

Figure 16. Comparisons of the time history plots of the upper cylinder for the problem of two cylinders moving towards each other: (**a**) C_D; and (**b**) C_L.

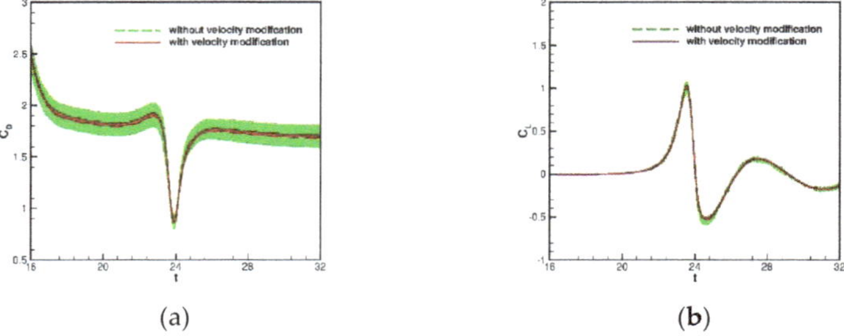

Figure 17. Comparisons of the effect of the velocity modification method for the problem of two cylinders moving towards each other: (**a**) C_D; and (**b**) C_L.

5.5. Free-Falling Circular Cylinder in Quiescent Flow

The free-falling circular cylinder problem will be simulated in this subsection. In a 2×6 rectangular box, a circular cylinder located at $(1,4)$ with a radius $r = 0.125$ is free falling from rest, as shown in Figure 18a. The densities of solid (ρ_s) and fluid $\left(\rho_f\right)$ are 1.25 and 1, while the non-dimensional fluid viscosity and gravity for the investigated problem are 0.1 and $(0, -980)$. The cylinder positions and vorticity contours at different times are plotted in Figure 18. In order to validate the present framework, comparisons of the time history of cylinder positions and velocity are also plotted at Figure 19. It can be clearly shown that the preset results obtained by the proposed framework and methodology with $\Delta x = \Delta y = 1/128$ and $\Delta t = 10^{-4}$ agree well with the reference benchmark results [29,30]. For the sake of completeness, C_D, C_L and ω are plotted in Figure 20a, while the effect of velocity modification method is plotted in Figure 20b. From Figure 20, it shows that the present framework can also well suppress for the free-falling cylinder problem. The pressure contours at t = 0.2 are also plotted to further show the benefit of employing velocity modification method. From Figure 21a, the smooth pressure is obtained, while Figure 21b clearly shows oscillated pressure near the downward cylinder interface.

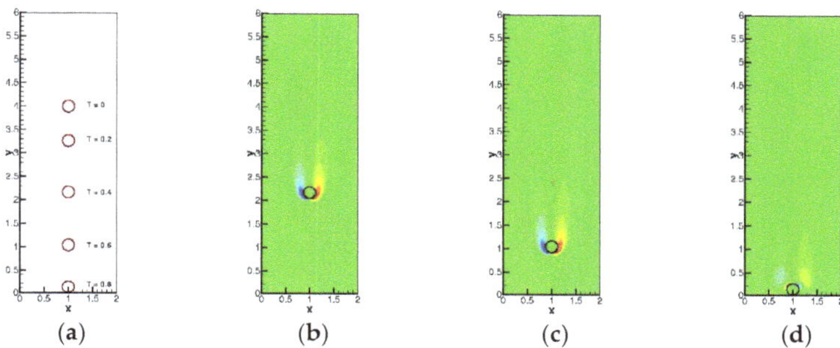

Figure 18. The predicted results for the free-falling circular cylinder problem: (**a**) the cylinder positions at different times; (**b**) vorticity contours at t = 0.4; (**c**) vorticity contours at t = 0.6; and (**d**) vorticity contours at t = 0.8.

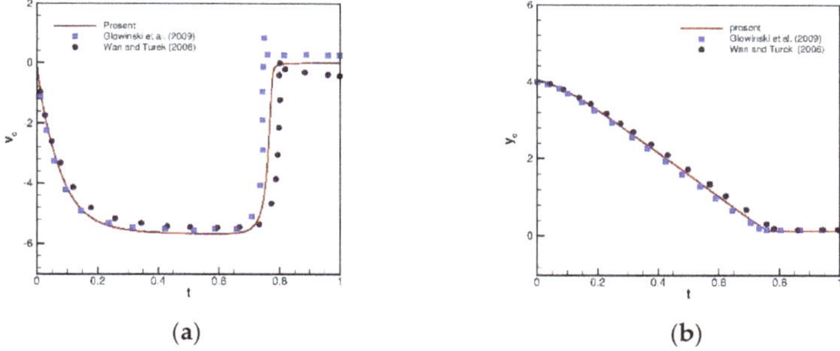

Figure 19. Comparisons of the results for the free-falling circular cylinder problem with benchmark results [29,30]: (**a**) velocity; and (**b**) centric y location.

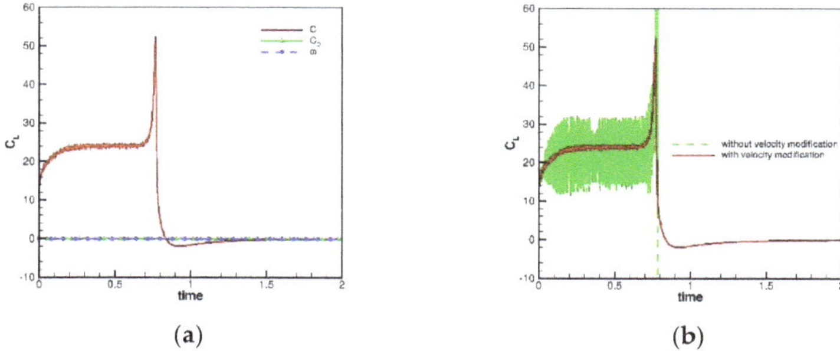

Figure 20. The time history plots for the free-falling cylinder problem: (**a**) C_D, C_L and ω; and (**b**) comparison of the velocity modification method for C_L.

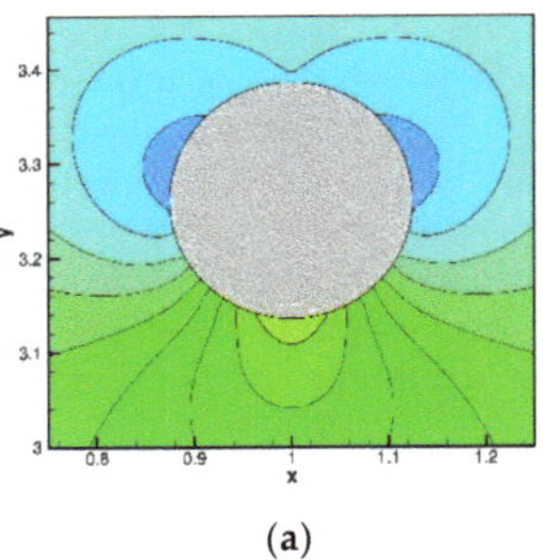 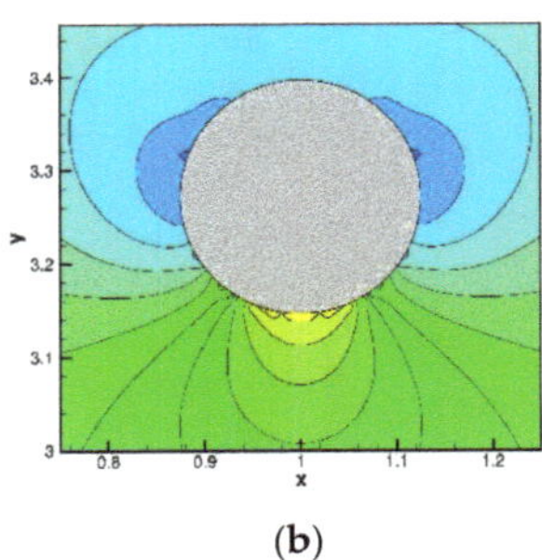

(**a**)   (**b**)

Figure 21. The predicted pressure contours at t = 0.2 for the free-falling cylinder problem: (**a**) with velocity modification method; and (**b**) without velocity modification method.

To show the advantage of the present framework, different density ratios (ρ_s/ρ_f), 1.05 and 1.01, were also investigated in this study. Figure 22 clearly shows that the present framework can correctly and consistently predict solid velocity and position. The applicability and efficiency of the present framework for solving the free-falling cylinder problem is thus confirmed.

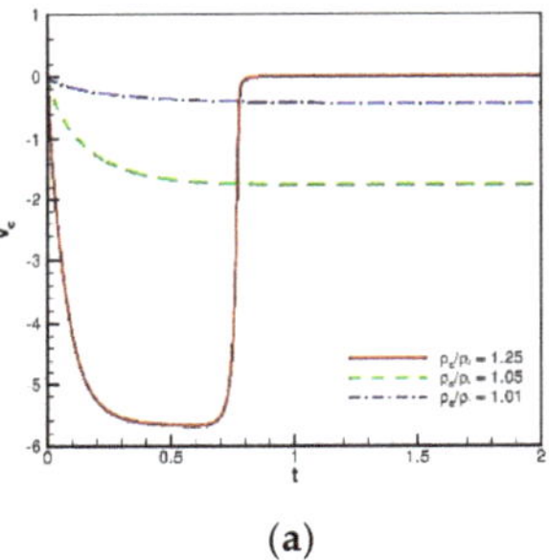 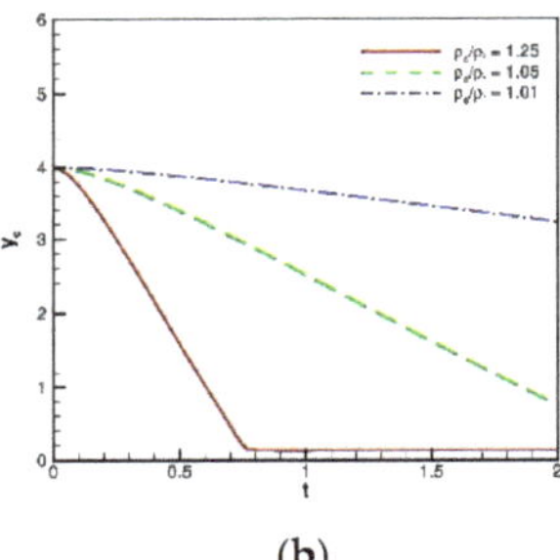

(**a**)   (**b**)

Figure 22. Comparisons of the results for the free-falling circular cylinder problem with benchmark results [25,26]: (**a**) velocity; and (**b**) centric y location.

5.6. Drafting–Kissing–Tumbling (DKT) Problem of Two Free-Falling Circular Cylinders in Quiescent Flow

Finally, the drafting–kissing–tumbling (DKT) problem was investigated in this subsection. In a computational domain of $(x,y) = (0 \sim 6, -1 \sim 1)$, there are two cylinders with a radius r = 0.125 located at (1, 0.001) and (1.5, −0.001). The density ratio and viscosity ratio was chosen as 1.5 and 0.01, while the gravity is (980, 0). To model the DKT scenario in this study, the repulsion force model [29] was employed in this study, with the choice of a stiffness parameter as 5×10^{-3} and effective force range as $3\Delta x$. Due to the large repulsion force, the sub-time stepping method is employed to prevent the divergence issue when solving the equations of motions. The idea is to introduce a sub-time step $\Delta t_s (= \Delta t/s)$, which is s times smaller than the original time step Δt, and solves the sub-time step s times. In practice, the fluid solver costs the most of the computational time, so the additional cost due to the sub-time stepping can be ignored. In this study, the sub-time step s is set as 20. The predicted pressure and vorticity contours with $\Delta x = \Delta y = 1/256$ and $\Delta t = 5 \times 10^{-5}$ at t = 0.15, 0.2, 0.25 and 0.3 are plotted in Figures 23 and 24. The comparisons of solid velocity and position with the referenced solution [2] are then made. From Figures 25 and 26, good agreements can be seen at $t \leq 0.17$, (drafting and first kissing stage). For $t > 0.17$ (the later kissing and tumbling stage), it still shows reasonable agreements. As indicated by [2], the discrepancy should be due to strong instability and the choice of collision parameter and the initial horizontal offset of two cylinders. It is noted that there is no drafting for the present framework if the horizontal offset is set to zero.

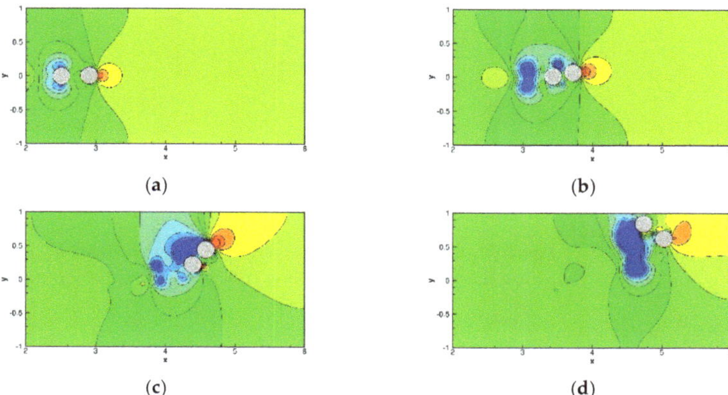

Figure 23. The predicted pressure contour plots for the drafting–kissing–tumbling (DKT) problem: (**a**) $t = 0.15$; (**b**) $t = 0.2$; (**c**) $t = 0.25$; and (**d**) $t = 0.3$.

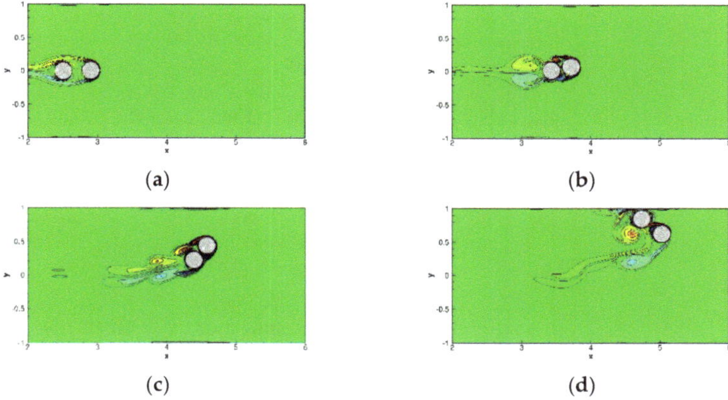

Figure 24. The predicted vorticity contour plots for the DKT problem: (**a**) $t = 0.15$; (**b**) $t = 0.2$; (**c**) $t = 0.25$; and (**d**) $t = 0.3$.

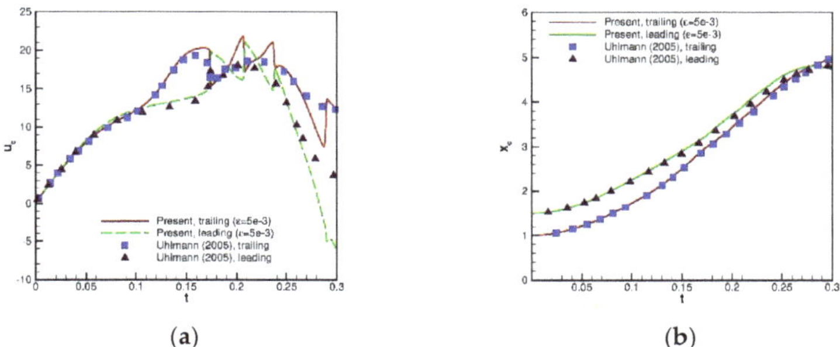

Figure 25. Comparisons of the x velocity (u_c) and x position (x_c) of two cylinders with referenced results for the drafting–kissing–tumbling (DKT) problems of two free-falling particles: (**a**) u_c; and (**b**) x_c.

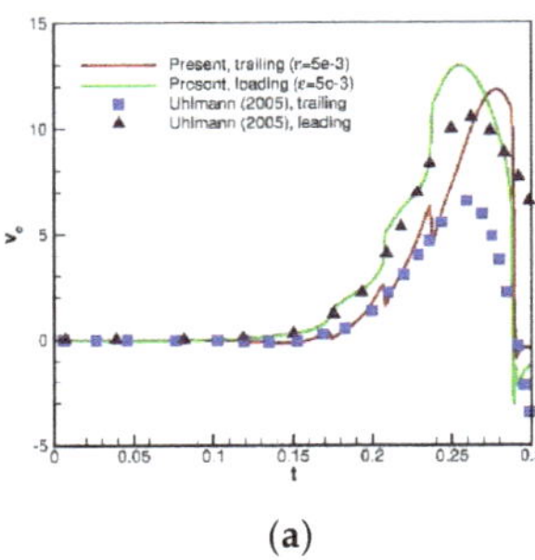 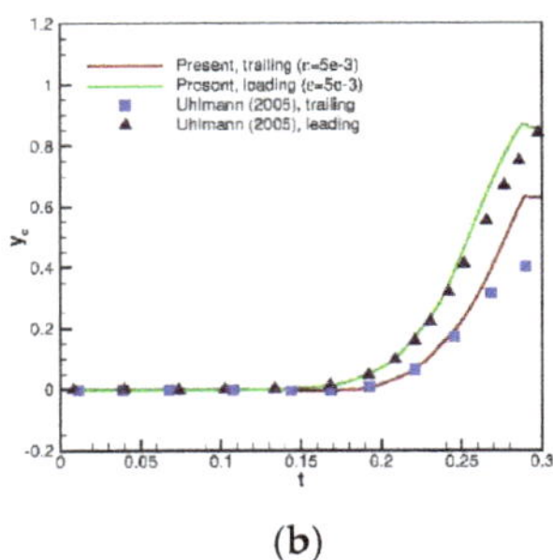

(a)   (b)

Figure 26. Comparisons of y velocity (v_c) and y position (y_c) of two cylinders with referenced results for the drafting–kissing–tumbling (DKT) problems of two free-falling particles. (**a**) v_c; (**b**) y_c .

6. Conclusions

In this study, we proposed a framework to model the flow–particle interaction problem. The solid object is modeled by the DIIB method, while the fluid velocity and pressure are solved by the IDFC^2 method. To prevent the instability problem of a density ratio close to 1, the hydrodynamic force and torque which appeared in the equations of motion are directly evaluated with modified interpolation kernel function. The methodology of evaluating the drag/lift forces and the treatment of a circular cylinder approaching the wall have also been addressed. To validate the applicability and accuracy, problems of one and two particles were investigated. To make the framework more stable, the idea of a sub-time stepping method was also employed for DKT problems of two particles. From the investigating validation/benchmarking problems, the simulated results reveal the applicability, accuracy and efficiency of the present framework.

Author Contributions: P.-H.C. developed the numerical model, performed the numerical simula-tions, analyzed the results, plotted the figures, as well as wrote the manuscript. H.C.W. analyzed the results and wrote/revised the manuscript. Y.Z.C. review and editing. R.B. review and editing. Y.-T.L. performed the numerical simulations, analyzed the results, as well as wrote the manuscript. All authors have read and agreed to the published version of the manuscript.

Funding: Yan-Ting Lin would like to thank the Ministry of Science and Technology, Republic of China for the funding support under project No. MOST 108-3116-F-042A-006.

Acknowledgments: Raymond Byrne acknowledge the Research Office at Dundalk Institute of Technology, Ireland.

Conflicts of Interest: The authors declare no conflict of interest.

References

1. Peskin, C.S. The immersed boundary method. *Acta Numer.* **2002**, *11*, 479–517. [CrossRef]
2. Uhlmann, M. An immersed boundary method with direct forcing for the simulation of particulate flows. *J. Comput. Phys.* **2005**, *209*, 448–476. [CrossRef]
3. Kempe, T.; Fröhlich, J. An improved immersed boundary method with direct forcing for the simulation of particle laden flows. *J. Comput. Phys.* **2012**, *231*, 3663–3684. [CrossRef]
4. Breugem, W.P.A. Second-order accurate immersed boundary method for fully resolved simulations of particle-laden flows. *J. Comput. Phys.* **2012**, *231*, 4469–4498. [CrossRef]
5. Zhang, L.T.; Gay, M. Immersed finite element method for fluid-structure interactions. *J. Fluid Struct.* **2007**, *23*, 839–857. [CrossRef]
6. Luo, K.; Wang, Z.; Fan, J. A modified immersed boundary method for simulations of fluid–particle interactions. *Comput. Methods Appl. Mech. Eng.* **2007**, *197*, 36–46. [CrossRef]
7. Wang, Z.; Fan, J.; Luo, K. Combined multi-direct forcing and immersed boundary method for simulating flows with moving particles. *Int. J. Multiph. Flow* **2008**, *34*, 283–302. [CrossRef]
8. Yang, J.; Stern, F. A Sharp Interface Direct Forcing Immersed Boundary Approach for Fully Resolved Simulations of Particulate Flows. *J. Fluids Eng.* **2014**, *136*, 040904. [CrossRef]

9. Horng, T.L.; Hsieh, P.W.; Yang, S.Y.; You, C.S. A simple direct-forcing immersed boundary projection method with prediction-correction for fluid-solid interaction problems. *Comput. Fluids* **2018**, *176*, 135–152. [CrossRef]

10. Wu, J.; Shu, C. Particulate Flow Simulation via a Boundary Condition-Enforced Immersed Boundary-Lattice-Boltzmann Scheme. *Commun. Comput. Phys.* **2010**, *7*, 793–812.

11. Hao, J.; Zhu, L. A lattice Boltzmann based implicit immersed boundary method for fluid–structure interaction. *Comput. Math. Appl.* **2010**, *59*, 185–193. [CrossRef]

12. Wang, L.; Guo, Z.; Shi, B.; Zheng, C. Evaluation of Three Lattice Boltzmann Models for Particulate Flows. *Commun. Comput. Phys.* **2013**, *13*, 1151–1172. [CrossRef]

13. Zhou, Q.; Fan, L.S. A second-order accurate immersed boundary-lattice Boltzmann method for particle-laden flows. *J. Comput. Phys.* **2014**, *268*, 269–301. [CrossRef]

14. Zhang, H.; Trias, F.S.; Oliva, A.; Yang, D.; Tan, Y.; Shu, S.; Sheng, Y. PIBM: Particulate immersed boundary method for fluid–particle interaction problems. *Powder Technol.* **2015**, *272*, 1–13. [CrossRef]

15. Coclite, A.; Ranaldo, S.; de Tullio, M.D.; Decuzzi, P.; Pascazio, G. Kinematic and dynamic forcing strategies for predicting the transport of inertial capsules via a combined lattice Boltzmann-Immersed Boundary method. *Comput. Fluids* **2019**, *180*, 41–53. [CrossRef]

16. Coclite, A.; Ranaldo, S.; Pascazio, G.; Tullio, M.D. de A Lattice Boltzmann dynamic-Immersed Boundary scheme for the transport of deformable inertial capsules in low-Re flows. *Comput. Math. Appl.* **2020**, *80*, 2860–2876. [CrossRef]

17. Chiu, P.H.; Poh, H.J. Development of an improved divergence-free-condition compensated coupled framework to solve flow problems with time-varying geometries. *Int. J. Numer. Meth. Fluids* **2021**, *93*, 44–70. [CrossRef]

18. Chiu, P.H.; Lin, R.K.; Sheu, T.W.H. A differentially interpolated direct forcing immersed boundary method for predicting incompressible Navier–Stokes equations in time-varying complex geometries. *J. Comput. Phys.* **2010**, *229*, 4476–4500. [CrossRef]

19. Yang, X.L.; Zhang, X.; Li, Z.L.; He, G.W. A smoothing technique for discrete delta functions with application to immersed boundary method in moving boundary simulations. *J. Comput. Phys.* **2009**, *228*, 7821–7836. [CrossRef]

20. Chiu, P.H. An improved divergence-free-condition compensated method for solving incompressible flows on collocated grids. *Comput. Fluids* **2018**, *162*, 39–54. [CrossRef]

21. Glowinski, R.; Guidoboni, G.; Pan, T.W. Wall-driven incompressible viscous flow in a two-dimensional semi-circular cavity. *J. Comput. Phys.* **2006**, *216*, 79–91. [CrossRef]

22. Ding, L.; Shi, W.; Luo, H.; Zheng, H. Investigation of incompressible flow within 1/2 circular cavity using lattice Boltzmann method. *Int. J. Numer. Meth. Fluids* **2009**, *60*, 919–936. [CrossRef]

23. Sheu, T.W.H.; Chiu, P.H. A divergence-free-condition compensated method for incompressible Navier-Stokes equations. *Comput. Methods Appl. Mech. Eng.* **2007**, *196*, 4479–4494. [CrossRef]

24. Meneghini, J.R.; Saltara, F.; Siqueira, C.L.R.; Ferrari, J.A. Numerical simulation of flow interference between two circular cylinders in tandem and side-by-side arrangements. *J. Fluids Struct.* **2001**, *15*, 327–350. [CrossRef]

25. Mahir, N.; Altac, Z. Numerical investigation of convective heat transfer in unsteady flow past two cylinders in tandem arrangements. *Int. J. Heat Fluid Flow* **2008**, *29*, 1309–1318. [CrossRef]

26. Dehkordi, B.G.; Moghaddam, H.S.; Jafari, H.H. Numerical simulation of flow over two circular cylinders in tandem arrangement. *J. Hydrodyn. Ser. B* **2011**, *23*, 114–126. [CrossRef]

27. Slaouti, A.; Stansby, P.K. Flow around two circular cylinders by the random vortex method. *J. Fluids Struct.* **1992**, *6*, 641–670. [CrossRef]

28. Xu, S.; Wang, Z.J. An immersed interface method for simulating the interaction of a fluid with moving boundaries. *J. Comput. Phys.* **2006**, *216*, 454–493. [CrossRef]

29. Glowinski, R.; Pan, T.W.; Hesla, T.I.; Joseph, D.D.; Périaux, J. A fictitious domain approach to the direct numerical simulation of incompressible viscous flow past moving rigid bodies: Application to particulate flow. *J. Comput. Phys.* **2009**, *169*, 363–426. [CrossRef]

30. Wan, D.; Turek, S. Direct numerical simulation of particulate flow via multigrid FEM techniques and the fictitious boundary method. *Int. J. Numer. Meth. Fluids* **2006**, *51*, 531–566. [CrossRef]

MDPI

St. Alban-Anlage 66

4052 Basel

Switzerland

Tel. +41 61 683 77 34

Fax +41 61 302 89 18

www.mdpi.com

Energies Editorial Office

E-mail: energies@mdpi.com

www.mdpi.com/journal/energies